RF-MEMS Technology for High-Performance Passives

The challenge of 5G mobile applications

RF-MEMS Technology for High-Performance Passives

The challenge of 5G mobile applications

Jacopo Iannacci
Center for Materials and Microsystems (CMM) Fondazione Bruno Kessler, (FBK) Trento, Italy

IOP Publishing, Bristol, UK

Media content is available from the book information online: https://doi.org/10.1088/978-0-7503-1545-6.

ISBN 978-0-7503-1545-6 (ebook)
ISBN 978-0-7503-1543-2 (print)
ISBN 978-0-7503-1544-9 (mobi)

DOI 10.1088/978-0-7503-1545-6

Version: 20171101

IOP Expanding Physics
ISSN 2053-2563 (online)
ISSN 2054-7315 (print)

British Library Cataloguing-in-Publication Data: A catalogue record for this book is available from the British Library.

Published by IOP Publishing, wholly owned by The Institute of Physics, London

IOP Publishing, Temple Circus, Temple Way, Bristol, BS1 6HG, UK

US Office: IOP Publishing, Inc., 190 North Independence Mall West, Suite 601, Philadelphia, PA 19106, USA

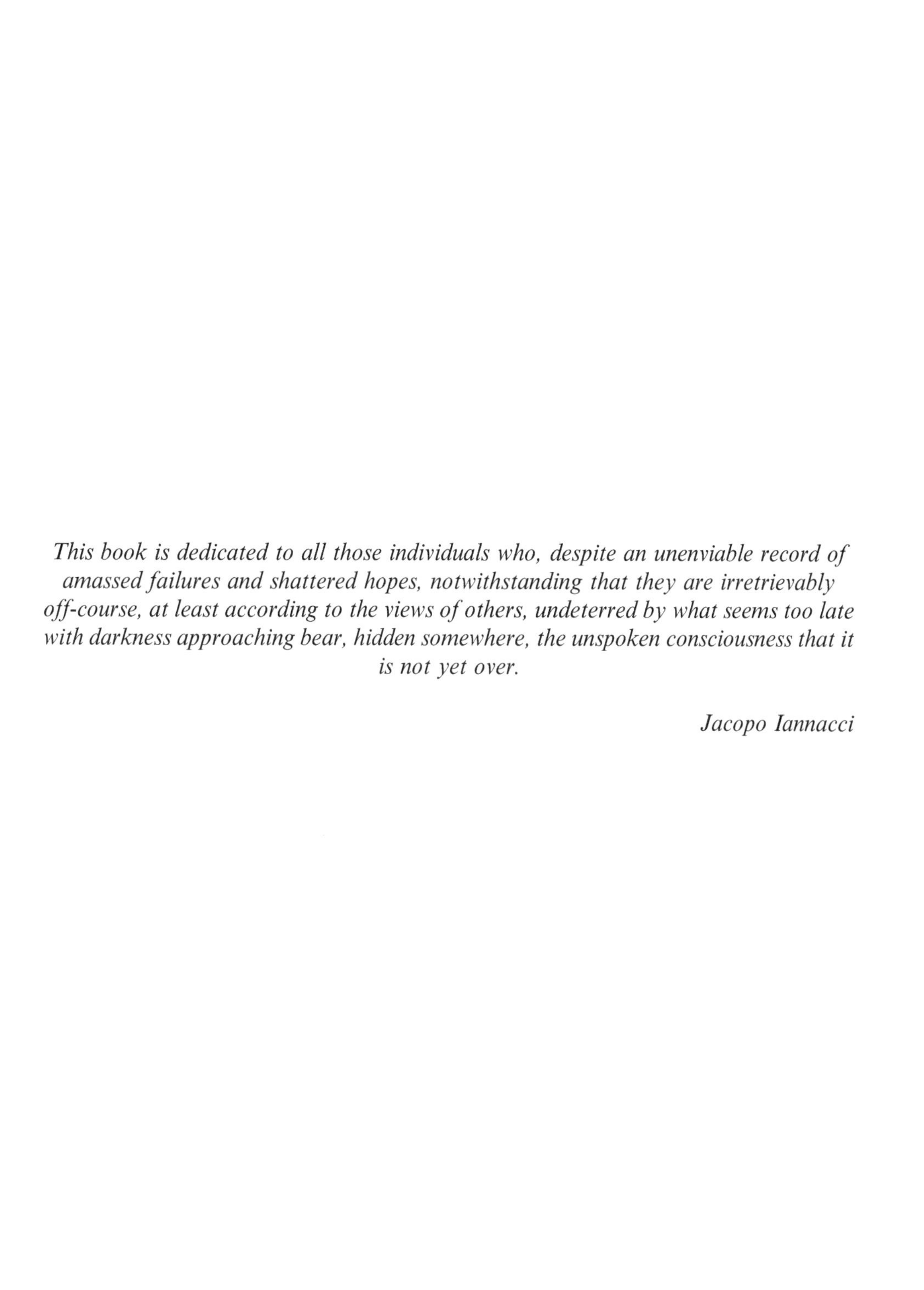

This book is dedicated to all those individuals who, despite an unenviable record of amassed failures and shattered hopes, notwithstanding that they are irretrievably off-course, at least according to the views of others, undeterred by what seems too late with darkness approaching bear, hidden somewhere, the unspoken consciousness that it is not yet over.

Jacopo Iannacci

Contents

Introduction

On the evening of June 24, 1982, the British Airways Flight 9, operated by the City of Edinburgh Boeing 747, was en route from Kuala Lumpur to Perth, in a multiple stopover haul from London (Heathrow), to Auckland. At around 20:40 (time of Jakarta) engine number 4 started malfunctioning and caught fire. According to standard drills, the crew shut it down and extinguished the flames. Shortly after, engine number 3 failed for no apparent reason, followed by engines 1 and 2. In a matter of two or three minutes, all four engines were down. It was the first time in the history of aviation that a Boeing 747 lost all four engines.

Starting from the altitude of the engines' failure, Flight 9 had approximately 23 minutes of gliding available before unavoidably touching down. Jakarta airport was within gliding distance, but a mountain in between had to be cleared; however, gaining altitude was not an option with no functioning engines. The only alternative was banking towards the open sea and trying an emergency ditching, even though, in the middle of the night and with no functioning engines at all, it was a very risky manoeuvre. On top of that, it would be the first time that a Boeing 747 would try to land on water.

After several minutes, just a step before the point of no return in which the crew would have had to aim the Boeing 747 nose to the Indian Ocean, engine 4 came back to life. Two minutes later, all four engines were back on track. Flight 9 gained altitude, cleared the mountain and landed safely in Jakarta.

The reason for the failure remained unknown until scientific investigations were carried out. The cause was found to be a cloud of volcanic ash produced by an eruption of Mount Galunggung, in West Java. The particles engulfed the engines, killing their power. It was only when the aircraft significantly lost altitude and the engines cooled down, that the ash solidified and broke off from the fans and rotating parts, allowing them to resume normal operation.

In summary, it was the first time a Boeing 747 had lost all four engines. The crew had absolutely no idea of the reason for such a failure, and found themselves faced with a critical decision: on the one hand, an emergency landing in Jakarta, impossible without engines, on the other, an emergency ditching in the Indian Ocean, at night, with no thrust, attempting a landing that had never been done before with a Boeing 747.

What Captain Eric Moody, First Officer Roger Greaves and Engineer Officer Barry Townley-Freeman kept doing in those endless minutes, was one, and only one thing. They repeated the engines' in-flight starting procedure, time after time, and even with no response at all, they tried again, dozens of times, relentlessly, without caring about the unmistakeable ineffectiveness and uselessness of their actions; without paying too much attention to the hopelessness of the whole situation.

While studying 'avoided' and, unfortunately, in other circumstances, unavoidable aviation accidents, I unexpectedly found an invaluable source of cases from which important lessons can be learned. For instance, what the crew of Flight 9 did in a matter of less than twenty minutes, projected on the unrolling of a lifespan—of course without the pressure of imminent disaster—could be the key to success. Pushing forth and trying to do what one feels like doing, against unfavourable circumstances and adverse and sceptical people, as well as, on top of it all, a lack of confirmation, might be a way to reach the target, because, primarily, it is the proof of your own motivation, belief and passion.

In fact, immediately following landing in Jakarta, Moody, Greaves and Townley-Freeman, instead of having champagne (and celebrating) with the cabin crew and passengers, looked through the booklet containing all the flight procedures and emergency drills dozens of times, because they were concerned that the engines' failure could have been their fault.

Now, if someone asks me to unfold the concept of commitment, I think that this example would sketch quite effectively the meaning that such a term bears.

Jacopo Iannacci

Preface

The scientific area of microsystems, known as MEMS (MicroElectroMechanical-Systems), has followed an evolutionary path intimately linked to that of semiconductors, albeit with relevant differentiation. MEMS and semiconductor technologies developed together, starting in the 1960s. While the manufacturing process of transistors was becoming more sophisticated, critical fabrication steps, like the patterning of buried conductive/piezoresistive thin-films and deposition of metals, began to be ventured to yield microstructures with mechanical properties. However, if transistors and semiconductors followed a relentless trend to miniaturization (*More Moore*) across the last four decades, microsystems developed increasing and diversifying transduction principles, aiming at broad functionalities in the fields of sensors and actuators (*More than Moore*).

Within the scenario of microsystem technologies, RF-MEMS, i.e. MEMS for Radio Frequency, started being investigated in the 1990s. Passive components for RF applications, such as waveguides, micro-relays, variable capacitors (varactors), as well as tunable filters, reconfigurable phase shifters, impedance tuners, and so on, were demonstrated in the literature. Their characteristics, in terms of low-loss, high-isolation and wide tunability, were outstanding if compared both to standard RF/microwave components and to semiconductors.

Such remarkable performances triggered, in the early 2000s, inflated expectations around the massive absorption of RF-MEMS components in mass-market applications, especially referring to mobile handsets and devices. In actuality, market forecasts were systematically disappointed for more than a decade, and the foreseen revolution never took place.

More recently, starting from 2014, RF-MEMS started to make their way into the market landscape, thanks to a need that other technologies were unable to address effectively. Modern 4th generation (4G) smartphones, due to the integration of more and more components, triggered a degradation trend in the quality of communication, which has been mitigated by the wide reconfigurability of MEMS-based RF passives. As a result, adaptive impedance tuners are now the first successful example of the exploitation of RF-MEMS technology in the consumer market segment. Besides, other MEMS components, such as high-performance switches, are starting to consolidate in applications like high-linearity RF Power Amplifiers (PAs) and RF Front Ends (RFFEs).

If this is the quite reassuring market snapshot of RF-MEMS today, further ahead lies the 5th generation (5G) of mobile communications and services, meant to be deployed approximately starting from 2020. The 5G standard will pursue the fundamental drivers of network densification and diversification. Compared to current 4G Long Term Evolution (4G-LTE), 5G will demand an increase of data volume up to 1000 times, of connected devices from 10 to 100 times, as well as reduced End to End (E2E) latency, Massive Machine Communication (MMC) and operation in the millimetre wave (mm-wave) range.

These harshly stringent system requirements will demand very-high performance, widely reconfigurable and frequency agile passive components. RF-MEMS technology holds the fundamental features to be a key enabling technology for 5G. From a different perspective, 5G could be the killer applicative scenario for massive consolidation of RF-MEMS in the mass-market landscape.

The aim of this book is to outline the outstanding intrinsic potential of RF-MEMS technology with reference to future 5G mobile communications and services. The way this objective is pursued is twofold. On one hand, the proper background concerning the characteristics of RF-MEMS is built, together with a definition of target specifications of interest for 5G. On the other hand, practical insight around the design and development of RF-MEMS passive components is provided by reviewing a few case studies of design concepts, also including multi-physics simulation approaches and techniques. In more detail, the book unfolds, chapter by chapter, as reported in the following.

Chapter 1 develops a comprehensive discussion on MEMS technologies. First, the inception of the concept of microsystems is analysed with reference to the evolution of semiconductor technologies, highlighting common features as well as their differentiation. The most common technology platforms for microsystems manufacturing are reported. Examples of consolidated (market) exploitations of MEMS sensors and actuators are also provided. Then, RF-MEMS are introduced, explaining their working principles and listing diverse actuation mechanisms. Subsequently, the most common categories of RF-MEMS devices are reviewed, focusing both on simple components, as well as on complex high-order reconfigurable networks.

Chapter 2 places the evolution of the RF-MEMS market, since the early days of the technology, under the spotlight. First, fundamental market analysis concepts, like the hype curve and technology push/market pull scenarios, are introduced. Then, boosted prospects around massive market penetration of RF-MEMS, counteracted, in fact, by fluctuations and disappointments for over a decade, will be studied in-depth. Two sets of reasons, namely intrinsic and extrinsic to RF-MEMS technology, are identified and discussed. Bearing in mind the lesson learned across multiple forecasted and then missed breakthroughs, a sound overview addressing the current state of the RF-MEMS market for mobile applications is provided.

Chapter 3 focuses on present and future mobile communication protocols. First, the working principles of mobile networks are introduced. The evolution of services and performance are reviewed, starting from the 1st generation (1G) of mobile communications, launched in the late 1970s, to the current 4th generation (4G). Subsequently, the frame of future 5th generation (5G) is depicted. The expected performance and services offered to end users are discussed. The main drivers of network virtualization, densification and diversification pushed forth by 5G will also be debated, covering challenges and key enabling technologies to make this vision real within a few years.

Chapter 4 deals with a few RF-MEMS design concepts, discussed through a quite practical and hands-on approach. First, capitalising on system-level requirements unrolled in the previous chapter, a few classes of passive components and a set of target specifications critical for 5G will be reported. Then, examples of RF-MEMS designs will be studied in detail, with the aid both of electromechanical and electromagnetic multiphysics simulations, as well as compact modelling techniques. With the aforementioned tools, case study designs will be altered, showing how this influences the RF-MEMS characteristics and how some can be tailored to 5G requirements.

Appendix A completes the book; showing the dynamics of a few RF-MEMS devices by means of experimental videos (acquired through optical microscopy techniques) and simulated animations.

Acknowledgements

The author would like to thank Dipl.-Ing. Harald Pötter, Dr.-Ing. Ivan Ndip and all the colleagues with the RF & Smart Sensor Systems Department at Fraunhofer Institute for Reliability and Microintegration IZM (Berlin, Germany), for providing access to the High Frequency Laboratory equipment and to the Ansys HFSS (High Frequency Structural Simulator) software tool, necessary to perform the experimental characterisation and Finite Element Method (FEM) simulations of the RF-MEMS devices discussed in chapter 4.

The author would also like to thank Prof. Dr. Gerhard Wachutka, Dr. Gabriele Schrag and Dr. Thomas Kuenzig with the Institute for Physics of Electrotechnology at Munich University of Technology TUM (Munich, Germany), for providing access to the characterisation laboratory and optical measurement facilities, discussed in appendix A.

Author biography

Jacopo Iannacci

Jacopo Iannacci is a Researcher in MEMS technology with the Center for Materials and Microsystems (CMM) at Fondazione Bruno Kessler (FBK) in Trento, Italy, since 2007. His scientific focuses are on (compact) modelling, design, optimization, integration, packaging and testing for reliability of RF-MEMS (Radio Frequency MicroElectroMechanical-Systems) passive devices/networks, EH-MEMS (Energy Harvesting MEMS), as well as SA-MEMS (MEMS Sensors and Actuators).

He received an MSc Degree in Electronic Engineering (2003) from the University of Bologna, Italy, and a PhD in Information Technology (2007) from the ARCES Center of Excellence at the University of Bologna, Italy, with focus on mixed-domain simulation and hybrid wafer-level packaging of RF-MEMS devices for wireless applications.

In 2005 and 2006, Jacopo Iannacci worked as a Visiting Researcher at the DIMES Technology Center of the Technical University of Delft, the Netherlands, where he focused on the development of packaging and integration solutions for RF-MEMS devices. In 2016, he visited the Fraunhofer Institute for Reliability and Microintegration IZM in Berlin, Germany, as a Seconded Researcher, studying the integration of RF-MEMS devices and their optimization against market requirements.

Jacopo Iannacci has authored and co-authored numerous scientific contributions for international journals and conference proceedings, as well as books and several book chapters in the field of MEMS and RF-MEMS technology.

IOP Publishing

Chapter 1

Introduction to MEMS and RF-MEMS: From the early days of microsystems to modern RF-MEMS passives

1.1 Introduction to semiconductor and microsystem technologies

Thinking of electronics and wondering about the intricacy of paths through which it has modified our habits, expectations and way of living in the last few decades, linking the invention of the transistor (by John Bardeen, Walter Brattain and William Shockley at Bell Labs in 1947) seems quite spontaneous, both to people holding technical skills in semiconductor technologies, as well as to the general public. It is unequivocal that the transistor, as an elemental building block of any electronic circuit, was and is still today, the key-element enabling the implementation of more complex and increasingly *smart* function/functionalities carried out by smaller, more integrated and less power-hungry devices.

Nonetheless, a quite critical consideration must be dragged under the spotlight before moving the discussion to the world of microsystems. In a rather effective attempt to reduce the complexity of a highly branched scenario, the transistor realizes two main functions, as it can be exploited as a relay, i.e. an ON/OFF switch, or as an amplifier, i.e. a device able to increase the amplitude of an electrical signal, according to a certain proportionality law. None of these functions were enabled or *invented* by the transistor. The first electrically operated switch, or relay, is attributed to the American scientist Joseph Henry in 1835. Its development was driven by the advancement of telegraph technologies. On the other hand, the first thermionic valve for amplification purposes was invented by John Ambrose Fleming in 1904. Bearing in mind this scenario, it is straightforward that transistors and, more in general, semiconductor technologies, have been playing a key role in the development of electronic devices for decades, with no leverage on the novelty or complexity of the function implemented by a single device with respect to its vacuum valve traditional counterpart. The actual key-enabling feature of semiconductor-based components is

doi:10.1088/978-0-7503-1545-6ch1

miniaturization, closely linked to the ease of integration. Just to provide a simple visual interpretation of the latter concept, it is sufficient to focus on the fact that the computational capacity of a modern smartphone, 55 or 60 years ago, would have required a medium-size apartment full of thermionic valves, relays, wires and power cables, to be implemented. As a matter of fact, miniaturization and integration enabled by semiconductor technologies, triggered a relentless trend in increasing the implemented complexity, counterbalanced by a moderate and, therefore, affordable spread in manufacturing and production costs, as well-framed by Moore's law [1]. As a matter of completeness, the latter states that with the advancement of technology, the number of transistors that can be integrated in a square inch of silicon doubles roughly every two years.

On the other hand, the development of microsystem technologies has followed a path that exhibits several factors in common with semiconductors, but is also different from such technologies for many other critical aspects.

Microsystems, which are universally referred to with the MEMS acronym (MicroElectroMechanical-Systems), are millimetre/sub-millimetre devices, realizing a certain transduction function between two (or more) distinct physical domains, among which the mechanical is always involved. More simply, regardless of the specific function it is conceived for, a MEMS device always features tiny structural parts that move, bend, stretch, deform and/or contact together. These peculiarities make microsystem devices particularly suitable for the realization of a very-wide variety of micro-sized sensors and actuators.

Provided with these basic concepts, a few considerations around the differences and similarities of MEMS versus semiconductor technologies can now be developed. Commencing from the most obvious diversities, while semiconductor devices are active, i.e. able to amplify an electrical signal, MEMS are exclusively passive, i.e. can just attenuate an electrical signal. However, transistors do not feature any movable or deformable part, i.e. they never exploit the mechanical/structural domain to realize transduction functions.

From a technological point of view, MEMS and semiconductors share most of the micro-fabrication steps, as will be discussed later, in more detail. Both feature selective deposition/removal of conductive/insulating thin-films by means of lithography, despite a few peculiar steps and sequences of fabrication that are typical of MEMS only.

Both MEMS and semiconductors pursue the concept of miniaturization. However, if semiconductor devices, beyond down-scaling, implement in the electrical/electronic domain the multi-physical function of traditional components, MEMS often miniaturize classical objects, keeping their transduction across physical domains. To this regard, the example of the aforementioned relay is quite explanatory. The traditional electrically-operated switch exploits the transduction between the electrical and mechanical domain to realize the ON/OFF function. The transistor (when exploited as a switch) realizes such a function entirely in the electrical/electronic domain. In contrast, a MEMS switch commutes between the ON/OFF state by coupling the mechanical and electrical domains, likewise the traditional device, despite the former being two or even three orders of magnitude smaller compared to the latter.

Also, importantly, the concept of miniaturization is inflected in a radically diverse fashion, when referring to semiconductor and microsystem (MEMS) devices. In the first case, as mentioned earlier, the trend in down-scaling has been continuous for decades. In order to build a more circumstanced idea, in Complementary Metal Oxide Semiconductor (CMOS) technology, the reference geometrical feature characterizing the transistor is the channel length. In the mid-1980s such a length was around 4 μm, in the mid-1990s it was roughly 600 nm, in 2010 it reached 30 nm, while nowadays it is well-below 20 nm [2]. This trend is, broadly speaking, addressed by the turn of phrase '*More Moore*', indicating the substantial hold of validity of Moore's law.

The concept of miniaturization *played* by microsystem technologies is completely different with respect to the aforementioned sketched scenario. First, there is no such thing as a trend in evolving technologies and processes in order to make the same MEMS device smaller, from one year to the next. Instead, a strong driver exists to implement more and more functionalities, possibly bringing them from the macro- to the micro-world. In other words, if the transistor was the same device over a number of decades, benefiting from being smaller and smaller as it becomes faster, less power consuming, more integrated and so on, the MEMS is a miniaturized object that benefits from implementing more and/or diverse sensing/actuating/transducing functions, by means of a device roughly the same size. Because of these characteristics, microsystems, as well as other non-standard technologies not mentioned here for brevity, are generally labelled by the turn of phrase '*More than Moore*' [3], indicating that their evolution through time does not follow Moore's law, as they cannot be standardized according to a development trend exclusively built upon the continual shrinking of dimensions.

Eventually, from a different perspective, the concept of miniaturization is radically dissimilar in quantitative terms, as well, when referring to semiconductors rather than microsystems. While CMOS transistors, as mentioned before, are framed today in the range of tens of nanometres, a MEMS sensor/actuator can span from a few micrometres (in-plane dimensions), to hundreds of micrometres, or even to a few millimetres. Therefore, if a MEMS switch is nearly invisible to the naked eye when compared to a traditional relay, it is massive when placed beside a CMOS transistor.

In the following subsections, a few key considerations will be developed around the early days of MEMS, the most diffused micro-fabrication techniques and their market applications. Such concepts will help one to understand the core topic of this work, which will be introduced immediately after, i.e. MEMS for Radio Frequency applications, universally known as RF-MEMS.

1.1.1 The genesis of MEMS

As already discussed, the non-standard underlying peculiarity of microsystems with respect to semiconductors has emerged. Due to this reason, development of MEMS as a whole did not follow a well-established path, making it difficult to determine an exact point in time corresponding to the conception of microsystems.

From the point of view of technology, key-fabrication steps developed together with the growth of semiconductor technologies starting in the 1950s. Nonetheless, exploitation of such techniques aimed towards the manufacturing of microsystems commenced later, in the early 1970s. The advancement of silicon-based semiconductor technology motivated the scientific community to investigate, beside critical aspects related to the electrical/electronic characteristics, the mechanical properties of the materials involved in the manufacturing of semiconductors. To this regard, significant contributions can be found, for instance, in the valuable work of several authors concerning the mechanical properties of both bulk materials [4] and deposited thin-layers [5–7], dating from the mid-1950s to mid-1960s. Nevertheless, the exploitation of such techniques aimed towards the manufacturing of micro-devices with movable parts and membranes emerged later, in the period from the second half of the 1970s to the beginning of the 1980s.

Examples of how to exploit anisotropic etching to obtain a variety of 3D suspended structures from a silicon substrate are provided in [8]. Such techniques, together with those typically exploited for the fabrication of transistors and Integrated Circuits (ICs), led to the realization of miniaturized pressure sensors [9], accelerometers [10,11], switches [12,13], and other devices for various applications, such as in the optical and biomedical fields. A remarkable article summarizing the state-of-the-art microsystem technologies, and providing a comprehensive outlook around diverse applications, was authored by Petersen [14] at the beginning of the 1980s.

Nonetheless, it was with the further maturation of the surface micromachining fabrication technique [15] that the development of microsystems started to receive a significant boost, leading to the concept of MEMS sensors and actuators, as we know it today. A relevant contribution is represented by the work of Howe and Muller [16] in 1983, in which micro-cantilevers and double supported beams were realized in polycrystalline silicon, and released suspended above the substrate via silicon oxide utilized as sacrificial layer. Since then, a wide variety of MEMS-based sensors, actuators, and various mechanisms, like gears and micro-motors, have been developed, tested, and reported in the literature [17–19].

In addition to what has just been outlined, another significant element of diversity emerges between semiconductor and microsystem technologies. The integration of transistors in silicon-based substrates on one hand, and the novelty of the electrical/electronic properties of semiconductor materials, on the other hand, stimulated, for decades, the aggregation and strengthening of a distinct and unprecedented multi-disciplinary domain of science. Within it, classical physics and chemistry converged, mainly driven by the development of technology, together with mathematics and electrical techniques, necessary for the functional understanding of novel devices. This resulted in boosting the development of the electronics discipline, which today encompasses, at the same time, computer-based codes for designing circuits and systems, as well as quantum theory to describe the physical behaviour of state-of-the-art semiconductor devices.

On a different plane, microsystems did not prosper as a well-structured discipline in the beginning. Essentially, the possibility to realize mechanical micro-devices was seen in the early days as a sort of side-branch of the more and more standardized semiconductors manufacturing stream. As the small scientific community working in the MEMS field was mainly involved in the development of transistors and active devices, a consolidated background in mechanics and structural mechanics was essentially missing. However, the community of mechanical/structural engineers was neither particularly interested, nor even fascinated, by the idea of bringing part of their expertise *down to* the micro-world, especially from the mid-1950s to the 1960s, when the massive development of modern aircraft was driving research. It is only much later, roughly from the first half of the 1990s, that MEMS started to emerge as a self-standing discipline, where basic knowledge of physics, chemistry, electronics and fabrication, started to be blended together with structural mechanics, electromechanics and functional reliability. Such a statement is corroborated by the fact that in those years the first books, scientific journals and international conferences explicitly focused on microsystems started to emerge, supported by the growth of a sectorial community of researchers, designers, engineers and developers.

1.1.2 Micro-fabrication technology platforms

The manufacturing of semiconductor components, like transistors, consists in a sequence of steps in which different doses of dopant materials are selectively implanted/diffused within a substrate (typically silicon) to obtain (locally) certain electrical properties. The same implantation/diffusion, or alternatively the digging of deep trenches, can be performed to enhance isolation and reduce the cross-talk between adjacent devices. In addition, conductive and insulating layers are selectively deposited/grown, or deposited/grown everywhere and then selectively removed, in order to redistribute the electrical signals, from the intrinsic devices to the external world. The most important steps in the manufacturing of ICs are ion implantation, diffusion, epitaxial growth, chemical vapour deposition/physical vapour deposition and their variations, wet and dry etching, sputtering, evaporation, and electrodeposition of metals. The selection of the areas that have to undergo one or more of the previously listed steps, is achieved by means of lithography [20]. The typical cross-section view of CMOS transistors is depicted by the schematic in figure 1.1.

Looking at the cross-section, a couple of considerations must be developed. First, the intrinsic device is *made into silicon*, where differently doped areas (with diverse electrical/electronic properties) are obtained through the aforementioned fabrication steps. Second, the transistors themselves are very small, while the metal and insulating layers stacked above them, necessary to redistribute the electrical signals from each transistor terminal to the external world, can be several times the size of the semiconductor intrinsic devices. In figure 1.1 the stack is completed at the top by a metal ball, necessary for chip-level integration, i.e. for mounting/interfacing the CMOS circuit into a more complex system or sub-system.

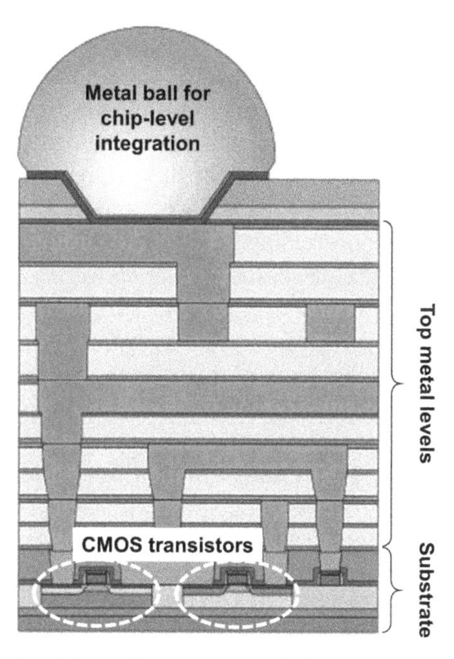

Figure 1.1. Schematic cross-section of CMOS transistors [21].

Microsystems, as mentioned earlier, are manufactured relying (for the most part) on the fabrication steps commonly utilized in the processing of semiconductors, even though they are arranged according to different sequences. Despite the fact that MEMS technology is not highly-standardized like CMOS, it is possible to identify two distinct processing sequences, which are the most diffused, both in the research and commercial production of microsystems. For the sake of completeness, there also exist highly-customized technology platforms strongly oriented to the manufacturing of MEMS. Such solutions enable one to achieve very-high aspect ratios and extremely precise geometrical features. Nonetheless, their pronounced customization makes them divert from CMOS-like processes, resulting in significantly higher costs and more articulated issues in terms of integration with other technologies. Because of their suitability for niche applications, these solutions will not be discussed further in this work. However, just to mention that the Lithography, Electroplating, and Moulding process (in German, Lithographie, Galvanoformung, Abformung—LIGA), is one of the best-known highly-customized technologies for the manufacturing of MEMS [22, 23].

Coming back to the most common aforementioned MEMS fabrication flows, they are substantially two: *surface micromachining* and *bulk micromachining*.

In surface micromachining processes, the substrate (silicon or other materials) is used as a sort of *ground floor* plane. All the selective deposition/removal of the conductive/insulating layers is performed above the substrate, through the techniques and steps mentioned above. From a conceptual point of view, a surface micromachining process is not crucially different from the above-CMOS stacking of layers depicted in figure 1.1. The substantial addition of a MEMS surface micromachining process is that membranes and movable parts need to be suspended in air. To do so, a temporary layer is necessary to support mechanically the micro-membranes during their manufacturing (e.g. through electrodeposition or sputtering). Afterwards, it has to be removed via an etching step, in order to release the so-called air-gaps, i.e. membranes suspended above the substrate and, therefore, free to move. Such temporary support is called a sacrificial layer, and it can be a photoresist material or a thin film deposited during processing [24, 25]. Accordingly, it can be stated that in surface micromachining MEMS devices are *made above silicon*.

A typical schematic cross-section of a MEMS surface micromachining process is reported in figure 1.2(a), while the microphotograph of physical gold-based MEMS devices manufactured with such a technology solution [26], is shown in figure 1.2 (b).

In this example, the silicon substrate is 625 μm-thick, air-gaps are around 3 μm, while the suspended gold membrane's thickness ranges between 2 μm and 5 μm. Each of the MEMS in figure 1.2(b) has in-plane dimensions of 2 mm by 0.7 mm.

In bulk micromachining processes, the substrate itself (typically, but not limited to, silicon) is exploited for the realization of the structural parts of MEMS. By means of performing selective etching (removal) of substrate specific regions, e.g. Tetramethylammonium Hydroxide (TMAH) based wet-etching [27] or Deep Reactive Ion Etching (DRIE) dry-etching, thin and deformable membranes are released [28]. Still keeping the same idiomatic expression as above, it can be stated that in bulk micromachining, MEMS devices are *made of silicon*. In light of the discussion developed up to this point, bulk micromachining of MEMS deviates from standard semiconductor processes more than surface micromachining. If the latter solution is, for simple MEMS structures, an extension of the above-CMOS stacking of layers, in the former it is silicon (and its mechanical properties) that is to be exploited as a structural material. It is straightforward that silicon mechanical structures of MEMS devices fabricated through bulk micromachining can be completed by selective deposition of conductive/insulating layers, in order to deploy proper electrodes and feeding lines to enable electromechanical transduction.

A typical schematic cross-section of a MEMS bulk micromachining process is reported in figure 1.3 (a), while a Scanning Electron Microscopy (SEM) image of a

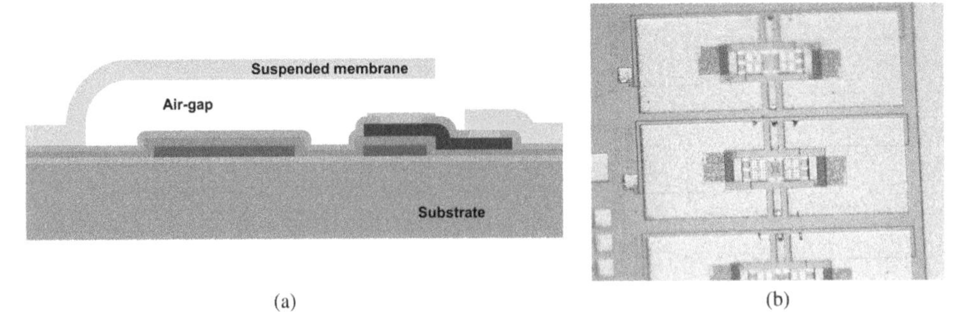

Figure 1.2. (a) Schematic cross-section of a typical surface micromachining MEMS process. (b) Microphotograph of physical MEMS devices realized by means of a surface micromachining process based on gold [26].

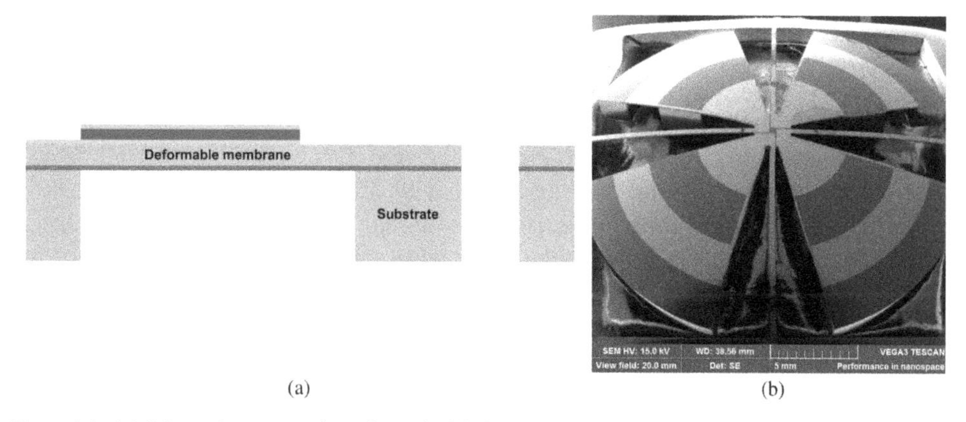

(a) (b)

Figure 1.3. (a) Schematic cross-section of a typical bulk micromachining MEMS process. (b) SEM image of physical EH-MEMS (MEMS for Energy Harvesting) devices realized by means of a bulk micromachining process based on silicon [29].

physical silicon-based EH-MEMS (MEMS for Energy Harvesting) realized with such a technology [29], is shown in figure 1.3 (b).

In the reported example, the initial silicon substrate is 390 μm-thick, while, after the etching step, the silicon deformable membranes are 20 μm-thick. The diameter of the suspended mechanical resonator in figure 1.3 (b) is 8 mm.

1.1.3 Applications of MEMS sensors and actuators

To complete the overview of microsystems' technologies on a general perspective, and before embarking on an in-depth discussion around RF-MEMS, a few references and examples of MEMS sensors' and actuators' market exploitations are going to be briefly provided.

In the first place, it should be highlighted that, if microsystem technologies started to develop in the 1970s (as discussed earlier), the first MEMS-based commercial product, namely, a surface micromachined accelerometer by Analog Devices [30], only showed up in the early 1990s, nearly two decades later. The rationale beneath the lengthy time to market characterizing the early exploitation of MEMS in the field of sensors and actuators, is a sizable mosaic composed of numerous tiles, featuring crucial pivots like reliability [31], packaging, integration, as well as costs and, last but not least, market readiness. These aspects will be treated with deeper specificity later, when discussing RF-MEMS.

Starting from the first commercial accelerometer mentioned in [30], MEMS-based sensors and actuators started to spread into diverse systems and devices, securing their presence in significant market segments. Accelerometers themselves, for instance, became, across the 1990s, a de facto standard in the automotive sector, being employed as deceleration sensors activating the inflation of airbags in the case of car crashes [32]. Concerning microsystem-based actuators, another successful exploitation of MEMS technology is related to micro-mirrors and, in particular, to their arrangement in high-density matrices of individual micro-devices. MEMS micro-mirrors have been commercially employed since the second half of the 1990s to form optical images on projectors' lenses, as well as on movie theatre screens. To this regard, Digital Micromirror Devices (DMDs) are well-known, and named in such a fashion because they are not controlled in an analogue way, but rather with two-state (ON/OFF) driving signals [33, 34].

In more recent years, MEMS accelerometers and gyroscopes, i.e. inertial devices sensitive to rotations around axes and to gravity, experienced a boost in terms of market volumes, significantly larger than traditional exploitation in the automotive sector. In the last decade, with the emergence of home video game consoles interacting with human motion, and later with the massive spread of smartphones and tablets, MEMS-based Inertial Measurement Units (IMUs) became standard components provided by a wide variety of Original Equipment Manufacturers (OEMs) [35].

Just to mention a few other classes of devices, MEMS are spreading in the switching/multiplexing of optical signals (actuators), miniaturized microphones (sensors) and, despite not being fully mature yet, loudspeakers (actuators), Energy Harvesters (EH) for environmental sources (sensors), pressure/gas/temperature sensors, strain gauges/deformation sensors, and so on. The targeted fields of applications are quite diverse, as they range from automotive to consumer electronics, as well as from space/defence to the medical/health sector.

Eventually, it can be stated with a certain confidence that trends in the exploitation of MEMS will keep ramping up in the years to come. The crucial paradigms of the Internet of Things (IoT) [36] and of the Internet of Everything (IoE) will demand the availability of smaller, cheaper, less power-hungry, multi-functional and more specialized sensors and actuators, to be integrated in increasing numbers within Smart-Cities, -Buildings, -Devices, -Factories, -Cars, -Objects, as well as the human body, e.g. through Body Area Networks (BANs).

1.2 Introduction to RF-MEMS

Bearing in mind the scenario described previously, the investigation of microsystem technologies for the realization of RF passive components is more recent. The first scientific contributions relating to the exploitation of MEMS-like technology steps for RF passives, started to appear on the landscape in the early 1990s, i.e. while MEMS accelerometers were establishing themselves as valuable commercial products. However, early examples of actual RF-MEMS devices only started to populate scientific literature in the second half of the 1990s.

At an early stage, miniaturization of microwave and millimetre-wave transmission lines and their implementation in micromachining technologies based on silicon, emerged as a quite promising research field [37] thanks to the outstanding performance figures in terms of low-loss and compactness, compared to traditional solutions [38]. The possibility of integrating fixed RF signal manipulation functions, e.g. through the realization of stubs [39], appeared as an additional strength of silicon-based waveguides. Among the various families of transmission line configurations available, and well-known for decades [40], microfabrication technologies are particularly suited for planar devices. Therefore, most of the attention and interest concerning their miniaturization was around the Coplanar Waveguide (CPW) and microstrip implementations of transmission lines.

Given these premises, fundamentals of the aforementioned waveguide configurations are going to be synthetically reported. A 3D schematic view of a waveguide in a coplanar (CPW) and microstrip configuration is shown in figure 1.4 (a) and figure 1.4(b), respectively. In the former, a central metallization acts as the RF signal lines, while two wider metallized patches are meant to be reference ground planes for the travelling RF signal. The central line and ground planes are separated by a gap, and all the metal layers lie on the same side of the substrate [41]. As the RF signal propagates along the waveguide, the electromagnetic field is confined between the central line and the ground planes, partially through the dielectric material underneath the metal layers, and partially through the air above them. Very often, a thin insulating layer is deposited on the substrate prior to the electroplating/evaporation of the CPW itself. This helps to reduce losses to the substrate. In the microstrip configuration, instead, the RF signal line is placed on top of the substrate, while a unique reference ground plane is metallized on the opposite face of the wafer. In this case, as the RF signal propagates along the waveguide, the electromagnetic field is mainly confined within the substrate, between the two metal layers (signal line and ground plane).

A basic lumped element network description useful to model the RF behaviour of CPWs and microstrip lines is proposed in [42] and depicted in figure 1.4 (c). Across

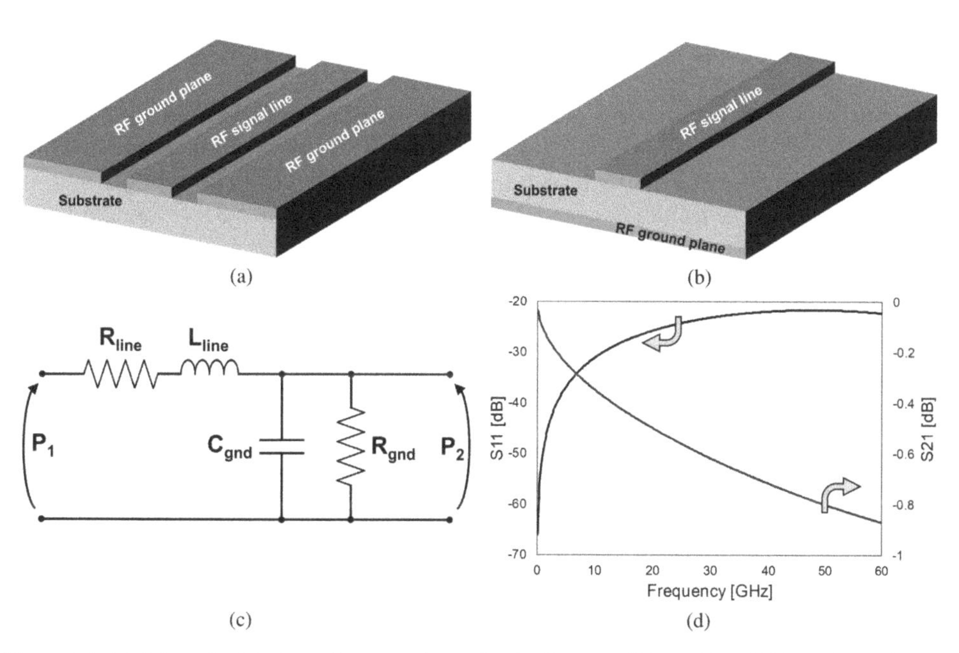

Figure 1.4. (a) Schematic 3D view of a transmission line in Coplanar Waveguide (CPW) configuration. (b) Schematic 3D view of a transmission line in a microstrip configuration. (c) Equivalent lumped element network of a CPW/microstrip line. P_1 and P_2 are the input/output terminations, respectively. (d) Typical S-parameters versus frequency characteristic of a CPW/microstrip line concerning reflection (S11 at P_1) and losses through the line (S21 at P_2).

the waveguide input and output ports P_1 and P_2, a resistance and inductance are inserted, representing the series resistive (R_{line}) and inductive (L_{line}) contributions of the metal RF line, respectively. On the other hand, the shunt-to-ground capacitance and resistance model the capacitive coupling (C_{gnd}) and the resistive losses (R_{gnd}) between the RF line and the ground plane/s, respectively, through the substrate material and through air. The value of the resistive and reactive components in figure 1.4 (c) are correlated to the physical properties of the transmission line, and can be parameterized in quite a straightforward fashion, in order to account for the most relevant geometrical features of the CPW or microstrip waveguide, like the length, gap, substrate thickness, and so on [43].

The RF behaviour of a typical CPW/microstrip line in terms of scattering parameters (S-parameters) versus frequency [42] is shown in figure 1.4 (d). The curves result from the simulation of a CPW by means of a Finite Element Method (FEM) software tool, in the frequency range from DC to 60 GHz. The S11 parameter indicates the fraction of RF signal reflected at the input port of the CPW; and as it is small on the whole frequency range (better than -22 dB), most of the RF signal flows into the waveguide. The S11 and S22 (reflection at the input and output ports, respectively) are particularly suited to provide an indication of matching between the characteristic impedance of the RF source and of the transmission line. Low values of S11/S22 mean good impedance matching, indeed. The S21 parameter indicates the amount of RF power reaching the output port of the CPW. Since its worst value (around -0.9 dB) is quite close to 0 dB (i.e. ideal zero losses), the attenuation of the RF signal introduced across the waveguide is limited overall across the analysed frequency span.

Besides the exploitation of a typical surface micromachining step like selective deposition of thin metal films, additional techniques started to be explored with the aim of improving the RF characteristics of miniaturized CPWs and microstrip lines. For instance, shallow tranches were etched in the gap between the RF signal line and the reference ground planes, in order to reduce the losses due to penetration of the electromagnetic field through the substrate, as reported in [44]. In other examples, bulk micromachining was used to remove most parts of the silicon substrate, yielding CPWs suspended above a thin membrane, resulting in a significant reduction of losses and parasitic coupling effects, as discussed in [45, 46]. Shortly after, MEMS technology began to be demonstrated for the realization of micro-switches [47] and variable capacitors (varactors) [48], as well as tunable filters [49], resonators [49] and programmable phase shifters [50], thus starting to address the crucial feature of reconfigurability. All these aspects are going to be discussed in more detail in the following subsections.

1.2.1 Switches and simple passives in RF-MEMS technology

Based upon the discussion developed in the above sections, it is straightforward that reconfigurability of a certain microsystem device can be enabled by inclusion of fabrication steps purposely conceived for such a target. Recalling what was mentioned at the beginning, in a case where the surface micromachining process is used, such a step is the exploitation of a sacrificial layer, meant to define and then release suspended structures. In contrast, when dealing with bulk micromachining, the MEMS structure has to be properly etched in order to be released and made free to move.

A few fundamental notions concerning actuation mechanisms are going to be recalled in the following pages. Subsequently, examples of physical RF-MEMS basic passive components will be discussed.

A brief review of actuation mechanisms

From a functional point of view, multi-physical coupling through which mechanical behaviour of movable RF-MEMS parts is controlled (and their characteristics reconfigured) can take place basically according to four different actuation principles: electrostatic, electromagnetic, piezoelectric, and thermoelectric [51]. These different mechanisms are going to be briefly explained.

► ***Electrostatic actuation***. Two electrodes, one fixed and one movable, are necessary, and they must face each other, as in a typical parallel plate capacitance configuration. When a voltage drop is applied across the two faces, the electrostatic attraction force makes the movable electrode approach the fixed one. Above a certain biasing threshold, called 'pull-in voltage', the movable part collapses onto the underlying fixed one. Figure 1.5 shows the schematic cross-section of an electrostatically controlled cantilever MEMS series ohmic switch [52].

In more detail, figure 1.5 (a) reports the cantilever switch in its rest position (no bias imposed). In the latter condition, the input/output terminals (T_1 and T_2) are disconnected, and the micro-relay is in the OFF state, as indicated by the switch symbol above the schematic. However, when a voltage V_{bias} larger than the pull-in threshold is imposed between the movable (Act_1) and fixed (Act_2) electrodes, the contact between T_1 and T_2 is closed, and the switch commutes to the ON state, as indicated in figure 1.5 (b).

► ***Electromagnetic actuation***. The suspended MEMS membrane has either to be made of or coated with a ferromagnetic material, in order to be sensitive to magnetic field variations. In addition, a magnetic field must be generated by driving a current across a coil, and the former has to surround the deformable membrane. In such a

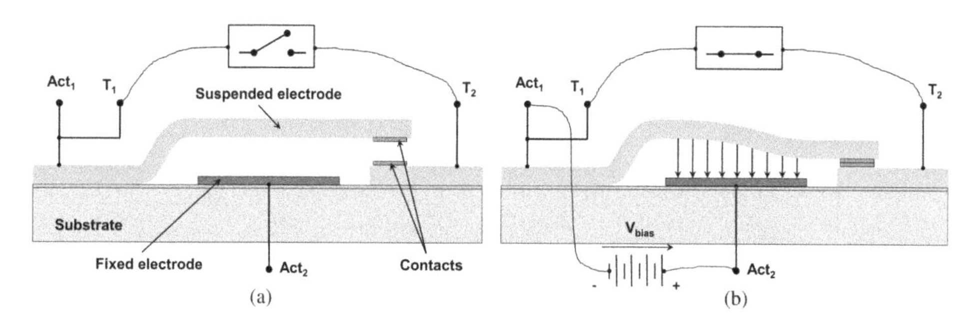

Figure 1.5. (a) Schematic cross-section of a cantilevered MEMS series ohmic switch controlled through the electrostatic principle, in the rest position (OFF state). (b) Schematic cross-section of the actuated or pulled-in position (ON state) when a bias voltage is imposed between the fixed and the floating electrode.

way, when the bias current is imposed, the MEMS part deforms due to the interaction between the magnetic-sensitive material and the external induced magnetic field [53]. The schematic cross-section of a cantilever series ohmic MEMS switch driven through electromagnetic actuation is reported in figure 1.6.

In particular, figure 1.6 (a) shows the schematic MEMS switch in the rest position, i.e. when no bias current is driven across the terminations Act_1 and Act_2. In this case, high-impedance is detected between the switch input and output ports, named T_1 and T_2, and the switch is OPEN (OFF state). On the other hand, when a current is driven through the coil, a magnetic field builds around the MEMS and the latter deforms until reaching pull-in, as shown in figure 1.6 (b). In such a circumstance, the impedance between T_1 and T_2 commutes to a very-low value, due to the physical contact between the two metal patches under the cantilever free end, and the switch is CLOSE (ON state).

► ***Piezoelectric actuation***. The suspended MEMS membrane must be covered/coated by a thin-film of material holding piezoelectric properties. As known, piezoelectric materials, which fundamentally behave electrically as insulators, exhibit the property of deforming/expanding when subjected to a voltage drop across their opposite faces [54]. As the piezoelectric thin-film is typically patterned above the MEMS structural part (made of gold, silver, copper, etc), its expansion due to the piezoelectric effect results in a downward (momentum induced) displacement [55]. The schematic cross-section of a cantilever series ohmic MEMS switch driven through piezoelectric actuation is shown in figure 1.7.

In the MEMS rest position, depicted in figure 1.7 (a), the switch is OPEN (OFF state). In contrast, when a bias voltage is imposed between Act_1 and Act_2, the piezoelectric material expands and induces commutation of the micro-relay to CLOSE condition (ON state) between T_1 and T_2, as shown schematically in figure 1.7 (b).

► ***Thermoelectric actuation***. In this case, the property of materials thermal expansion is exploited to drive the MEMS movable part/s. An electrical current is driven across the suspended membrane that heats up due to its resistance and, therefore, expands because of the temperature increase [56]. Alternatively, thermal

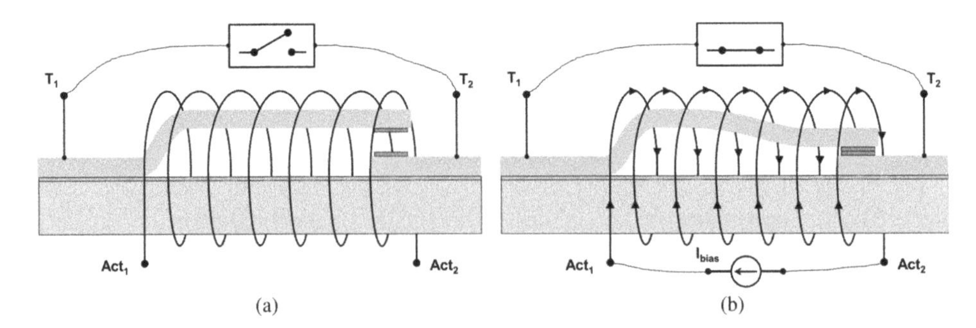

Figure 1.6. (a) Schematic cross-section of a cantilevered MEMS series ohmic switch controlled through the electromagnetic principle, in the rest position (OFF state). (b) Schematic cross-section of the actuated or pulled-in position (ON state) when a bias current is driven through the coil that induces a magnetic field around the movable MEMS membrane (coated with a ferroelectric material).

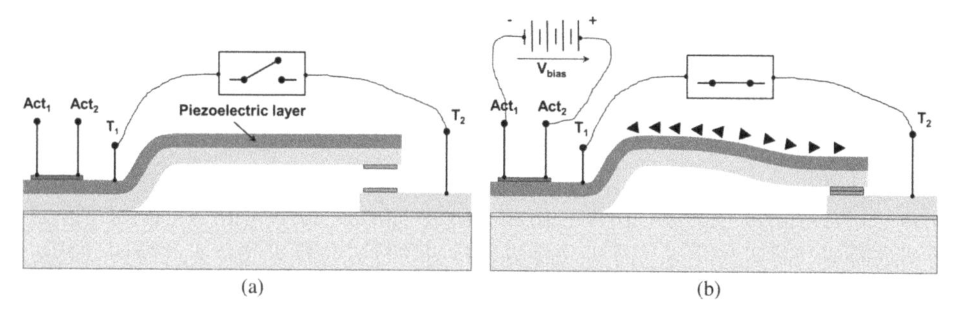

Figure 1.7. (a) Schematic cross-section of a cantilevered MEMS series ohmic switch controlled through the piezoelectric principle, in the rest position (OFF state). (b) Schematic cross-section of the actuated or pulled-in position (ON state) when the piezoelectric film is subjected to a bias voltage.

expansion of the suspended membrane can also be induced generating the heat not directly into the MEMS part itself, but, for instance, embedding micro-heaters underneath the device. This latter solution makes it possible to use materials with much higher resistivity (like polycrystalline silicon) than metals typically employed for a microsystem's structural parts (like gold, copper, aluminium, etc). Therefore, it is sufficient to drive a fairly low current through the heater in order to obtain the desired increase of temperature [57, 58]. The schematic cross-section of a cantilever series ohmic MEMS switch driven through thermoelectric actuation is shown in figure 1.8.

In more detail, figure 1.8 (a) reports the switch in the rest, i.e. OPEN, position (OFF state). In contrast, when a bias voltage is imposed across Act_1 and Act_2, a current is driven through the MEMS, and the heat generated brings commutation of the micro-relay in the CLOSE position (ON state), shown schematically in figure 1.8 (b).

Among these reviewed mechanisms, electrostatic actuation is certainly the most commonly used for controlling RF-MEMS devices. There are multiple motivations for this choice. One reason being that, at the technology level, electrostatic actuation does not require the deposition of exotic materials, e.g. with piezoelectric or ferromagnetic properties, therefore easing the manufacturing process and also limiting the costs. Furthermore, at the operation level, this kind of driving method does not induce irreversible changes in the mechanical properties of the MEMS, that may happen, for instance, with thermoelectric actuation. Additionally, the effort in terms of energy employed to control MEMS devices is lower if compared with other methods, among which thermoelectric and electromagnetic actuations are definitely the most energy consuming. As in electrostatically controlled MEMS, the physical contact between the movable and fixed electrodes must be avoided to prevent a short-circuit, so that no current flows through the device, leading to virtual zero-power consumption of the micro-relay, in both ON/OFF configurations. In fact, small current leakages are always present along the DC biasing lines. Nonetheless, they lead in any case to very limited amounts of power necessary to drive the MEMS. In addition, the electrostatic actuation of MEMS will be often referenced in the practical examples shown later in this work. In light of all these considerations, further technical discussion around electrostatics is briefly reported below.

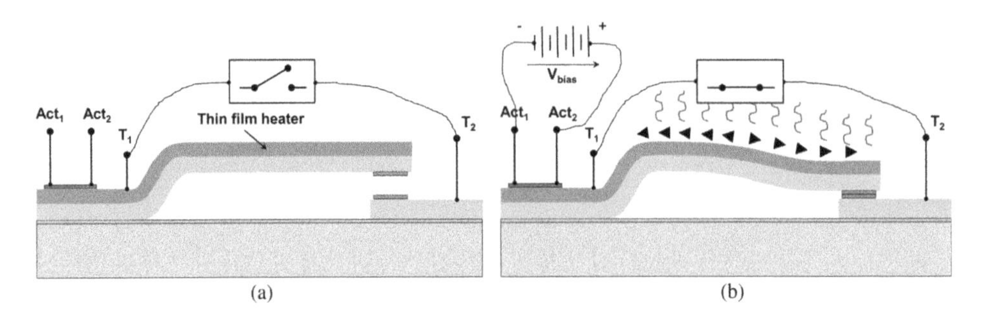

Figure 1.8. (a) Schematic cross-section of a cantilevered MEMS series ohmic switch controlled through the thermoelectric principle, in the rest position (OFF state). (b) Schematic cross-section of the actuated or pulled-in position (ON state) when a bias current is driven across the suspended membrane, causing the heating and the subsequent thermal expansion of the MEMS structure.

An electrostatically controlled MEMS device can be effectively represented as a parallel plate capacitor, with one fixed and one movable plate, as discussed in [59] and reported in figure 1.9.

The one Degree Of Freedom (1 DOF) parallel plate schematic is shown in figure 1.9 (a). The lower plate is mechanically constrained, while the upper one is joined to a mechanical spring whose elastic constant is k. The spring allows displacement along the vertical x-axis. The area of both plates is A, while their initial distance is x_0. They are immersed in air (dielectric constant ε_{air}) and the effects of gravity on the upper plate and spring can be neglected.

When a biasing voltage V_b is imposed across the two plates, and when it is below the pull-in threshold ($V_b < V_{PI}$), the situation is reported in figure 1.9 (b). Due to electrostatic interaction, the upper plate moves downward of x_1. The electrostatic attraction force F_{el} is expressed by equation (1.1).

$$F_{el} = \frac{1}{2}\frac{\varepsilon_{air}A}{(x_0 - x_1)^2}V_b^2 = \frac{1}{2}\frac{C(x)V_b^2}{(x_0 - x_1)} \tag{1.1}$$

At the same time, still referring to figure 1.9 (b), the spring elongates of x_1, giving rise to the mechanical restoring force F_{mech}, opposed to F_{el}, expressed by Hooke's law as follows:

$$F_{mech} = -kx_1 \tag{1.2}$$

Such a situation refers to an equilibrium condition since, given a certain $V_b < V_{PI}$, F_{mech} counteracts F_{el}, and the movable plate remains steadily at a distance equal to $x_0 - x_1$ from the underlying fixed electrode. Nonetheless, it should be noted that, while F_{mech} linearly depends on the distance between the plates, commonly referred to as the air-gap, F_{el} depends on its square value. By solving the system of equations (1.1) and (1.2) with respect to the voltage, it is possible to derive the pull-in voltage (V_{PI}), expressed as follows:

$$V_{PI} = \sqrt{\frac{8}{27}\frac{kx_0^3}{\varepsilon_{air}A}} \tag{1.3}$$

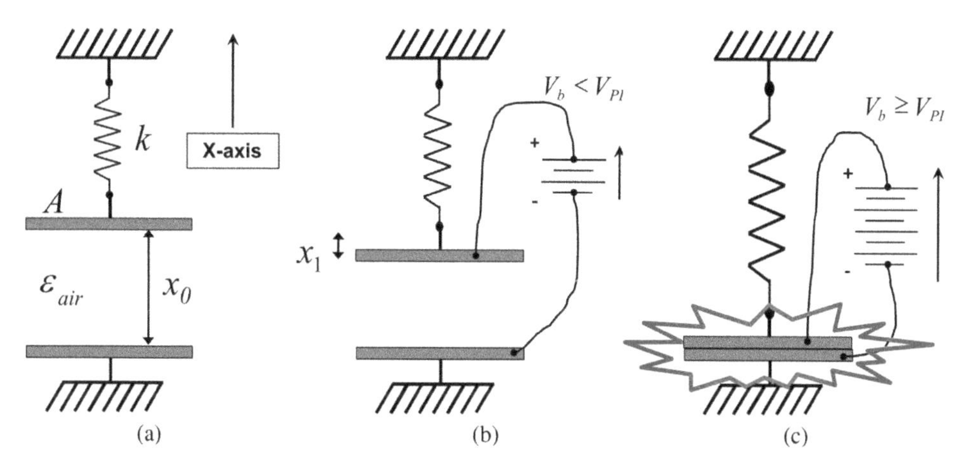

Figure 1.9. (a) Schematization of an electrostatically controlled MEMS device as a parallel plate capacitor, with one plate fixed and the other movable, when no bias voltage is applied ($V_b = 0$). (b) Displaced movable plate when a bias voltage lower than pull-in threshold ($V_b < V_{PI}$) is imposed across the two plates. (c) Collapsed (pulled-in) movable plate when a bias voltage equal or larger than pull-in threshold ($V_b \geqslant V_{PI}$) is imposed across the two plates.

Such a threshold bias level makes the movable plate collapse abruptly above the underlying fixed plate, as reported in figure 1.9 (c). In other words, the vertical displacement of the upper plate towards the lower one can be controlled in an analogue fashion as long as $V_b < V_{PI}$, corresponding to one-third of the initial air gap (x_0). From a physical point of view, V_{PI} is the limiting level for which F_{el} becomes too large to be counteracted by F_{mech}, and the system becomes unstable. As figure 1.9 (c) is a simplified schematic, no insulating layer is indicated between the two collapsed plates. In fact, physical contact must be avoided when pull-in occurs, as it would short-circuit the two plates. For this reason, in real MEMS a thin insulating layer (of thickness t_{ins}) is always deposited above the underlying fixed electrode. Alternatively, or in conjunction with such a layer, electrically floating stoppers (i.e. elevated posts) can also be deployed, in order to allow clearance between the pulled-in electrodes. Once pull-in takes place, if the bias voltage is progressively decreased, there exists another threshold value corresponding to the release (detaching) of the collapsed plate, named pull-out voltage (V_{PO}), which is expressed as follows:

$$V_{PO} = t_{ins}\sqrt{\frac{2kx_0}{\varepsilon_{ins}A}} \tag{1.4}$$

As mentioned before, since F_{el} depends on the square of the imposed voltage, V_{PO} is typically much smaller than V_{PI}. This means that the pull-in/pull-out (actuation/release) characteristic of an electrostatically controlled MEMS device exhibits a certain hysteresis, as it clearly emerges from the plot in figure 1.10.

When decreasing the bias voltage, right after pull-out occurs, the vertical displacement characteristic reconnects with the one previously obtained for increasing V_b levels (of course lower than V_{PI}). Eventually, since F_{el} depends on the square

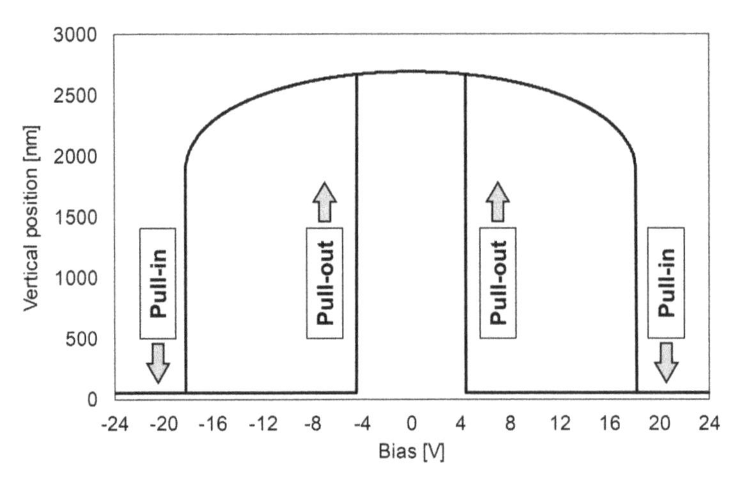

Figure 1.10. Typical pull-in/pull-out characteristic, i.e. vertical displacement versus imposed bias, of an electrostatically controlled MEMS device.

of the imposed voltage as indicated by equation (1.1), the pull-in/pull-out characteristic does not change if positive or negative bias levels are applied. This characteristic is straightforward when looking at the plot in figure 1.10, which is symmetric with respect to the 0 V vertical axis.

Next, examples of physical RF-MEMS simple components discussed in the literature are going to be reported. For the sake of homogeneity in explanation, all the photographs of the physical samples shown refer to electrostatically controlled RF-MEMS devices realized in the same surface micromachining technology platform, detailed in [60, 61]. In all of the following microphotographs, the MEMS devices and CPWs/microstrip lines are made of electrodeposited gold, with a thickness ranging between 2 μm and 5 μm. The thickness of the sacrificial layer and, therefore, of air-gaps, is always around 3 μm. A buried polycrystalline silicon layer (high-resistivity) is exploited to realize fixed DC biasing electrodes, underneath suspended MEMS membranes. Moreover, an additional aluminium buried layer (low-resistivity) is employed to implement RF signal underpasses and contact areas. A schematic cross-section of this technology was previously reported in figure 1.2 (a). Nonetheless, examples referring to other technologies discussed in the literature will also be reported.

RF-MEMS ohmic and capacitive switches

Briefly summarizing what was discussed earlier, microfabrication technologies enable the manufacturing of miniaturized waveguides, mainly in CPW and microstrip configuration. By adding MEMS specific fabrication steps, it is possible to realize suspended thin membranes, which can be driven/controlled according to different transduction mechanisms. The focus of this work is on electrostatically controlled RF-MEMS. Therefore, most of the examples that are going to be discussed in the following will refer to such a type of multiphysics coupling.

Having said that, the fundamental building block that enables the reconfigurability of RF-MEMS is represented by the micro-relay (or switch). The switching function, despite being based on a 2-state (OPEN/CLOSE) configuration, can be implemented according to different fashions, as it can be ohmic or capacitive, as well as series or shunt [62]. Of course, these listed features can be paired together according to all of the possible combinations. As representative examples, series ohmic and shunt capacitive RF-MEMS switches are going to be discussed in detail.

Figure 1.11 shows the microphotograph of a cantilevered RF-MEMS series ohmic switch electrostatically controlled, in CPW configuration.

The movable membrane is placed in-line on the RF signal path. It is anchored on one side and free to move on the other, as shown in the close-up in figure 1.11 (b). As a contact area is placed underneath the suspended tip, metal-to-metal contact is established when the MEMS is pulled-in. Therefore, the switch is OPEN when the MEMS is OFF (rest position), while it is CLOSE when the MEMS is ON (pulled-in), as reported in the simplified schematics in figure 1.12 (a) and figure 1.12 (b), respectively. The cantilever exhibits measured pull-in and pull-out voltages of 65 V and 50 V, respectively. The experimental RF behaviour in terms of the S-parameters is also reported in figure 1.12, and refers to a frequency range from DC up to 40 GHz. In particular, figure 1.12 (c) shows the reflection (S11) and isolation (S21) when the MEMS micro-relay is OFF, while figure 1.12 (d) reports the reflection (S11) and loss (S21) when the MEMS micro-relay is ON.

When OPEN, most of the power is reflected, as indicated by S11 in figure 1.12 (c) that ranges between ~0 dB and −0.6 dB. Isolation (S21) is better than −25 dB up to 10 GHz, and better than −10 dB up to 40 GHz. The S21 worsens as frequency increases, because parasitic series resistance is present between the suspended MEMS cantilever tip and the underlying contact area. When in the CLOSE state, the S11 is better than −25 dB up to 40 GHz, indicating quite good impedance

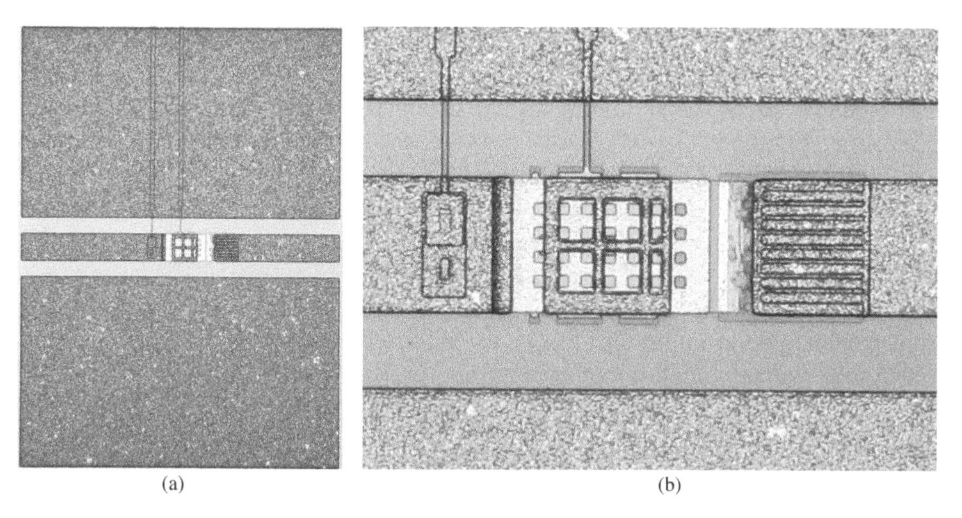

Figure 1.11. (a) Microphotograph of an RF-MEMS cantilevered series ohmic switch in CPW configuration. (b) Close-up of the micro-relay. The suspended cantilever is 180 μm long and 100 μm wide.

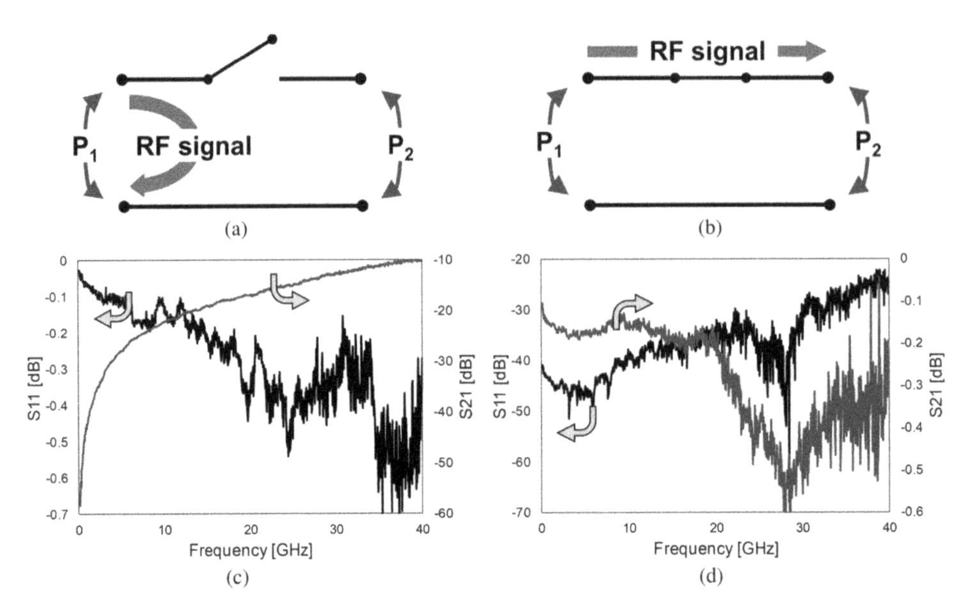

Figure 1.12. (a) Simplified schematic of the series ohmic micro-relay in figure 1.11 in the OPEN state (MEMS switch OFF). (b) Simplified schematic of the micro-relay in the CLOSE state (MEMS switch ON). (c) Measured reflection (S11) and isolation (S21) of the micro-relay in the OPEN state (MEMS switch OFF) from DC up to 40 GHz. (d) Measured reflection (S11) and loss (S21) of the micro-relay in the CLOSE state (MEMS switch ON) from DC up to 40 GHz.

matching between the RF source and the MEMS device (see figure 1.12 (d)). On the other hand, loss (S21) is very limited, being better than −0.5 dB up to 40 GHz. This result indicates the good quality of the ohmic contact between the pulled-in MEMS and the underlying metal area.

Figure 1.13 shows the microphotograph of an RF-MEMS shunt capacitive switch in a CPW configuration. In this case, the MEMS micro-relay is a membrane hinged at both ends (clamped-clamped configuration) to the ground planes, placed transversally across the RF line.

The behaviour of the shunt capacitive switch is dual with respect to the series ohmic one. First, no ohmic contact is established between the MEMS movable membrane and the RF underpass in any of the ON/OFF configurations, but, instead, by a 2-state capacitor. Then, the capacitance realizes a variable impedance path to the RF ground, rather than between the input and output terminations [62]. More details are reported in figure 1.14. When the MEMS is OFF, distance between the floating capacitance plate and the underlying one is maximum, therefore the shunt capacitance is minimum (C_{min}), as shown in figure 1.14 (a). Such a small capacitance realizes a high-impedance path to ground that lets most parts of the RF signal travel between the input and output of the device (CLOSE switch). On the other hand, when the MEMS is ON (pulled-in), the shunt capacitance reaches the maximum value (C_{max}), as shown in figure 1.14 (b). In this case, the large capacitance establishes a low-impedance path to ground that diverts most parts of the RF signal travelling across the device, shorting it to ground (OPEN switch). The

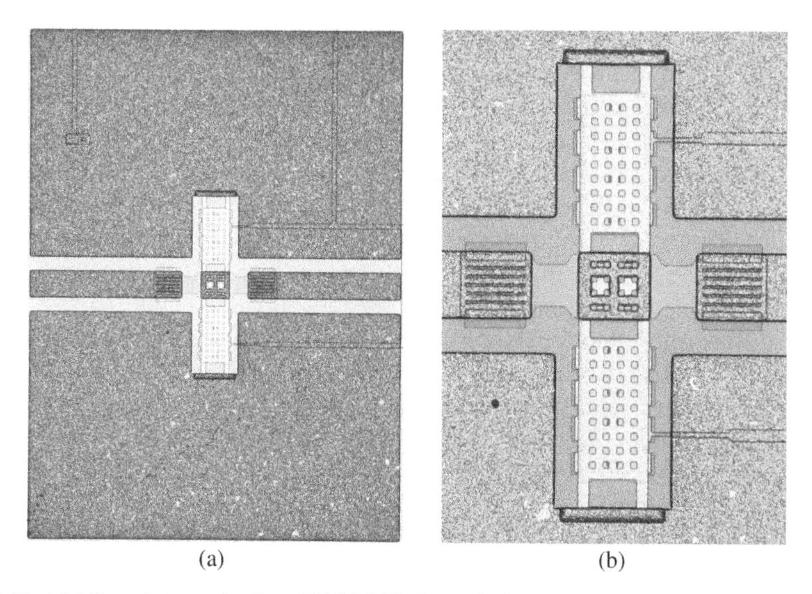

Figure 1.13. (a) Microphotograph of an RF-MEMS clamped–clamped shunt capacitive switch in a CPW configuration. (b) Close-up of the micro-relay. The suspended double-hinged membrane is 180 μm long and 100 μm wide.

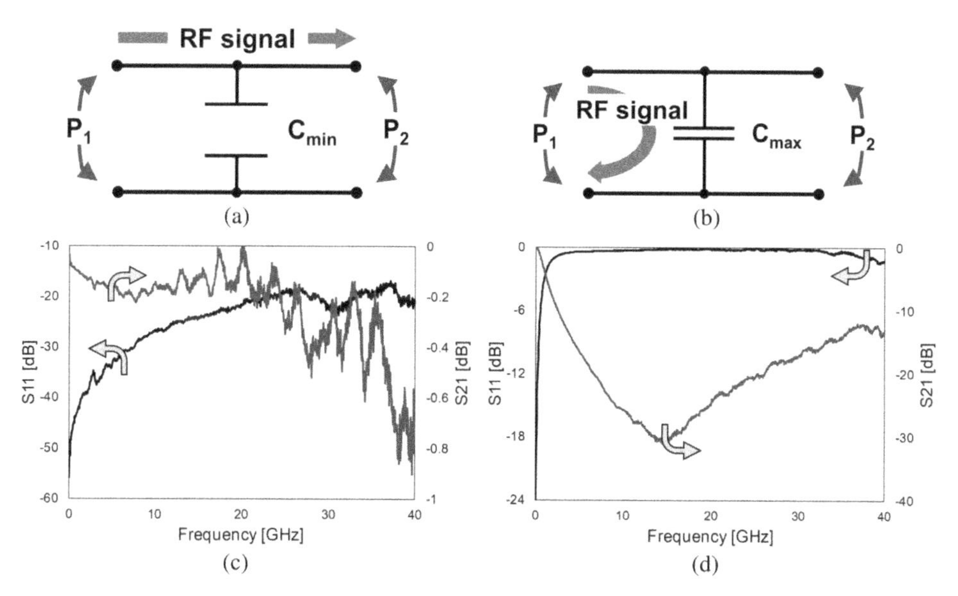

Figure 1.14. (a) Simplified schematic of the shunt capacitive micro-relay in figure 1.13 in the CLOSE state (MEMS switch OFF). (b) Simplified schematic of the micro-relay in the OPEN state (MEMS switch ON). (c) Measured reflection (S11) and loss (S21) of the micro-relay in the CLOSE state (MEMS switch OFF) from DC up to 40 GHz. (d) Measured reflection (S11) and isolation (S21) of the micro-relay in the OPEN state (MEMS switch ON) from DC up to 40 GHz.

measured S-parameters of the shunt capacitive switch in figure 1.13 are reported in figure 1.14, referring to the range from DC up to 40 GHz.

In particular, the S-parameters for the MEMS in the OFF state (CLOSE switch) are reported in figure 1.14 (c). Reflection (S11) is better than –20 dB up to 40 GHz, proving good impedance matching of the device. Moreover, loss (S21) is better than –0.8 dB up to 40 GHz. In contrast to the ohmic switch previously discussed, the S21 for the CLOSE switch exhibits increasing loss in the higher portion of the range. This is due to the shunt capacitance that, despite low (C_{min}), causes shorting to ground of a small part of the signal. In any case, having loss better than –1 dB up to 40 GHz is a very-good performance result. Figure 1.14 (d) shows the S-parameters for the MEMS in the ON state (OPEN switch). Starting from a few GHz, most parts of the RF signal are reflected, as indicated by the S11 curve. On the other hand, isolation (S21) exhibits its best value, i.e. –30 dB, around 15 GHz. It is straightforward that capacitive switches, in contrast to ohmic micro-relays, are significantly influenced by the resonant behaviour of the reactive elements [61, 62]. Therefore, they cannot exhibit remarkable performance on a very-wide band. Nonetheless, by exploiting their resonant characteristic, they can be optimized to outperform ohmic devices in very-well defined frequency ranges.

As mentioned above, micro-relays are fundamental bricks enabling the pronounced reconfigurability of RF-MEMS technology. In light of this consideration, micro-switches have always been widely discussed in the literature, since the early days of RF-MEMS, with respect to various aspects of their design, topology, RF and electromechanical characteristics, reliability, power handling, and so on. For instance, the work reported in [63] discusses an RF-MEMS switching unit with pronounced long-term operability and cycling up to 1 billion in the K-band (18–27 GHz). Moreover, the switch concept discussed in [64] refers to an in-package commercial RF-MEMS device, suitable for 4G mobile applications. On the other hand, [65, 66] discuss novel realizations of RF-MEMS switches with improved performance. Concerning pull-in voltage reduction, designs exhibiting actuation levels as low as 5–7 V were demonstrated in the literature [67].

RF-MEMS variable capacitors (varactors)

From the conceptual point of view, an RF-MEMS variable capacitor (commonly referred to as varactor) is not significantly different from a capacitive switch. Therefore, when the variable capacitance is inserted in shunt-to-ground configuration, its behaviour is well-described by the simplified circuit schematics previously shown in figure 1.14 (a) and figure 1.14 (b). Bearing in mind the discussion previously developed around the pull-in effect in electrostatically controlled MEMS, the varactor can be tuned in an analogue way (i.e. continuously) just from the rest position (zero bias) to the pull-in threshold, i.e. ranging across one-third of the overall air-gap. After pull-in, the capacitance will abruptly commute to the maximum value.

The microphotograph in figure 1.15 (a) shows an RF-MEMS varactor based on an electrostatically controlled floating gold electrode, kept suspended by four meander-shaped flexible beams.

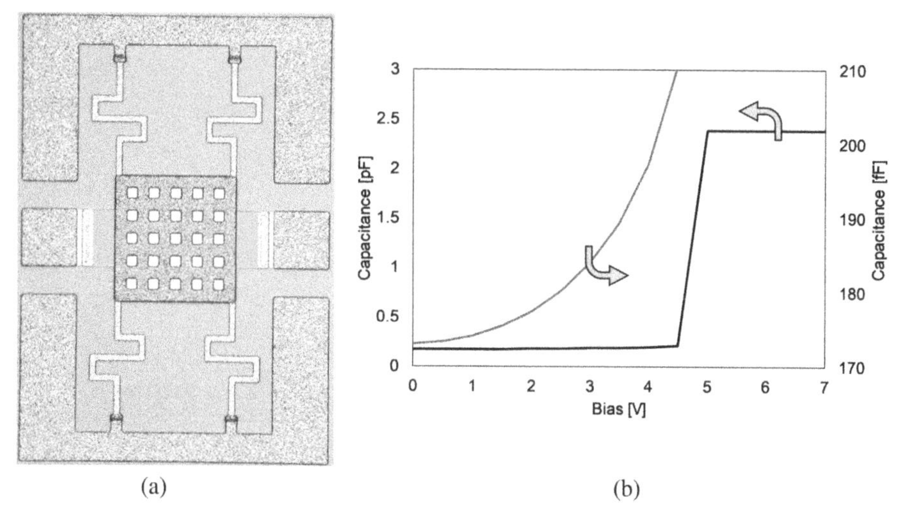

Figure 1.15. (a) Microphotograph of an RF-MEMS clamped-clamped shunt variable capacitor (varactor) in a CPW configuration. (b) Measured capacitance versus bias voltage (*C–V*) characteristic. The capacitance can be controlled in an analogue fashion until pull-in threshold is reached.

A fixed counter-electrode, lying beneath the movable MEMS, realizes the second capacitor plate [67]. The plot reported in figure 1.15 (b) shows the measured capacitance versus bias voltage (*C–V*) characteristic. Before pull-in, the capacitance ranges between ~170 fF and 210 fF (right vertical axis), and can be controlled continuously. When pulled-in, instead, the capacitance steps abruptly to ~2.4 pF (left vertical axis). Therefore, the ratio between the ON and OFF state capacitance (C_{on}/C_{off}) of the varactor in figure 1.15 is around 14, i.e. larger than one order of magnitude.

As for RF-MEMS switches, the scientific literature has always been populated by several contributions regarding varactors. To this purpose, an alternative design enabling a low-voltage controlled varactor is discussed in [68]. On the other hand, focusing on the extension of the varactor tuning range, the solution reported in [69] exploits the built-in intrinsic stress of the MEMS constitutive material, while [70] proposes a solution at the design level featuring a double actuation mechanism. Again, concerning the linearity improvement in RF-MEMS varactors response, the work discussed in [71] exploits a double varactor with anti-bias control in order to improve such a characteristic. On a different theme, on the side of medium-/long-term reliability, a double DC biasing pulse is reported in [72] with the aim of decreasing the amount of charge entrapped in the insulating layer and, in turn, the so-called voltage screening (i.e. drifting of the pull-in/pull-out characteristic due to charge injection and dipoles orientation within the insulators).

RF-MEMS (variable) inductors

Given the discussion developed up to now, it is straightforward that RF-MEMS technology is also suitable for the realization of high-performance inductors. The possibility to obtain suspended coils, as through a surface micromachining process

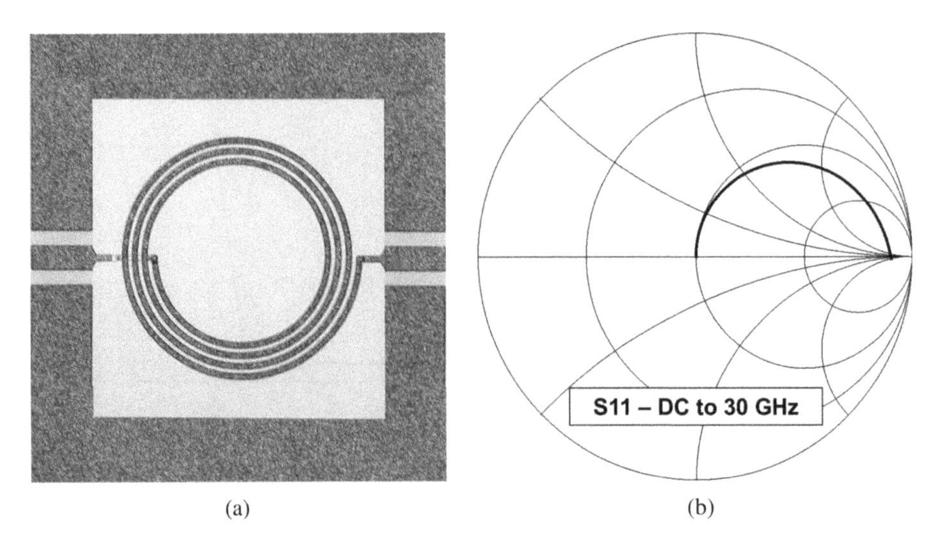

Figure 1.16. (a) Microphotograph of an RF-MEMS suspended coil inductor in a CPW configuration. (b) Smith chart of the input characteristic impedance (S11) from DC to 30 GHz.

featuring a sacrificial layer, as well as to manufacture metal lines above a thin substrate, as can be performed through etching the substrate from the back side (bulk micromachining), leads to a significant reduction of parasitic effects and, therefore, to an increase of the quality factor (Q-factor). The microphotograph in figure 1.16 (a) depicts a suspended coil inductor in a CPW configuration. The plot in figure 1.16 (b) shows the characteristic impedance of the inductor (S11) on a Smith chart, with reference to the frequency range from DC to 30 GHz.

The characteristic is always predominantly inductive, as confirmed by the trace mainly rotating in the upper half of the Smith chart. Another example of an air suspended inductor, based on a different coil design, is reported in [73]. Despite the fact that the characteristic of tunability is not as critical for inductors as it is for capacitors, different approaches to tune inductance, triggered by the flexibility of RF-MEMS technology, have been reported in the literature. A fairly popular approach to enable the tunability of RF-MEMS inductors is the exploitation of suspended coils non-planarity induced by residual stress within the patterned material. This leads to out-of-plane (i.e. vertical) displacement of adjacent coil turns, causing an inductance decrease if compared to the case of planarity. However, by driving a DC current through the coil, the heating induces a release of the intrinsic stress that temporarily improves the planarity and, therefore, increases the inductance value. Similarly, it is possible starting from a planar device in the rest position, to induce out-of-plane deformation by means of a bias current that reduces the inductance. Both these solutions are discussed in [74] and [75], respectively. A more exotic method to achieve inductor tunability is reported in [76]. In this case, a liquid is injected in the core, thus modifying its permeability and, therefore, the overall inductance value. More information and examples concerning high-performance (tunable) RF-MEMS inductors are developed in [77].

1.2.2 Complex reconfigurable passives in RF-MEMS technology

The discussion developed in the previous subsection outlined the most diffused classes of basic components realized in RF-MEMS technology. The common denominator of such groups is the implementation of a basic ON/OFF switching function (achievable in various ways), and/or of an actuation function that yields continuous (analogue) tunability within a certain range. Capitalising on the aforementioned basic components, and duplicating or combining them according to certain criteria within a unique physical device, it is possible to realize more complex RF-MEMS passive devices and networks, able to implement manipulation/treatment functions of RF/microwave/millimetre-wave signals, with highly-pronounced tunability/reconfigurability. In the following sections, the most common classes of complex RF-MEMS networks will be reported.

RF-MEMS switching units and matrices

Starting from the most elemental RF-MEMS component, i.e. the micro-relay, its duplication and arrangement within a unique device enables the expansion of the switching function from an ON/OFF configuration between one input and output termination, to multiple input and output branches. The microphotograph reported in figure 1.17 (a) shows an RF-MEMS Single Pole Double Throw (SPDT) in a CPW configuration. In summary, it is a T-type switch with one input and two output terminations, each of the latter being controlled by an electrostatically driven RF-MEMS series ohmic switch. Depending on the ON/OFF configuration of the two (independently controllable) micro-relays, the input signal can be driven to each of the two outputs, to both or none of them.

The plot in figure 1.17 (b) shows the measured S-parameter behaviour of the SPDT in the frequency range from DC to 40 GHz, and refers to a configuration in which one output channel is ON (CLOSE switch; pulled-in) while the other is OFF

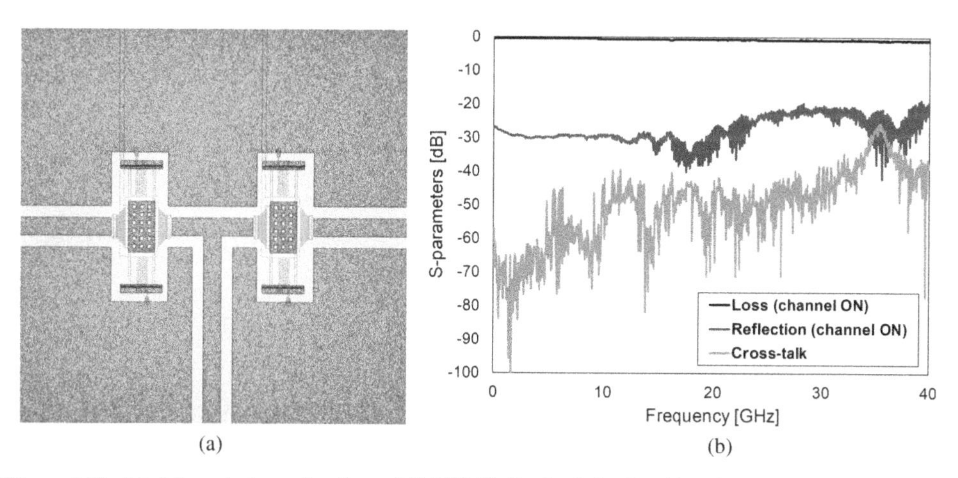

Figure 1.17. (a) Microphotograph of an RF-MEMS Single Pole Double Throw (SPDT) in a CPW configuration. (b) Measured reflection, loss and isolation between adjacent channels (i.e. cross-talk), from DC to 40 GHz.

(OPEN switch; rest position). The loss (S21) between the input and the conducting output branch is better than −1.1 dB and reflection (S11) is better than −22 dB to 40 GHz. Moreover, the isolation between the two output channels (S23), commonly addressed as cross-talk, ranges between −75 dB and −27 dB, and is always better than -45 dB to 35 GHz.

The complexity of the switching unit can be increased, thus extending the order of the realized function. Starting from the configuration in figure 1.17, the number of output branches can be enlarged, leading, for instance, to Single Pole Four Throw (SP4T) switching units, in which there are four output terminations, as well as, more in general, to Single Pole Multiple Throw (SPMT) configurations. In addition, commutation can be performed across multiple inputs and outputs, leading to actual switching matrices (e.g. 2 × 2, 4 × 4, or $N \times N$) that can be effectively designed and manufactured in RF-MEMS technology. Several examples are reported in the literature, proving the achievement of remarkable characteristics, both for simpler switching units like SP4T [78–80], as well as for various order switching matrices [81–84].

RF-MEMS tunable filters

Following the same approach, i.e. adding switching functionality, it is possible to enable tunability for other classes of passive devices, as in the case of RF filters. For instance, the hairpin filter configuration is well-known in the RF engineering community, and commonly exploited to shape a certain bandpass characteristic between the input and output terminations [85]. Such a class of filters was made tunable by implementing it in RF-MEMS technology, as discussed in [86] and depicted by the microphotograph in figure 1.18 (a).

The hairpin filter exploits the inductive contribution of U-shaped bends and their capacitive coupling to shape the bandpass response. In the RF-MEMS hairpin filter (microstrip configuration) in figure 1.18, the length of each U-shaped element can be increased by adding a patch selected by cantilevered series ohmic switches, similar to the device previously reported in figure 1.11 (a). Such a variation of geometry modifies the impedance of each U-shaped element, leading to a shift of the passed

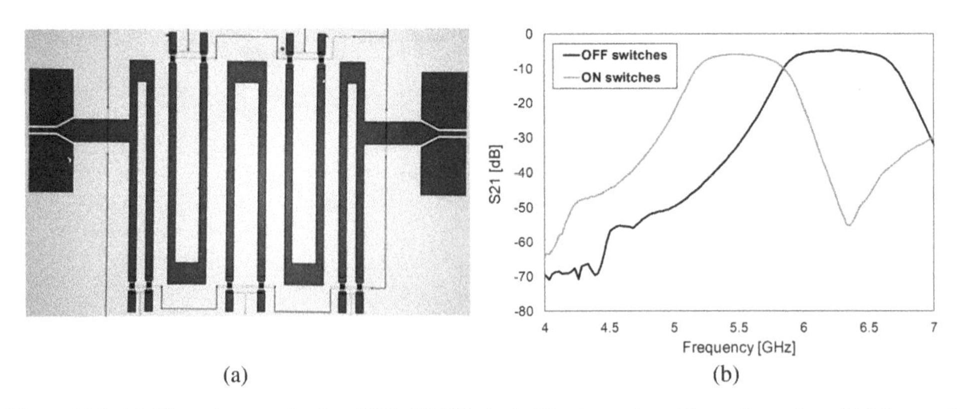

Figure 1.18. (a) Microphotograph of an RF-MEMS hairpin filter in a microstrip configuration. (b) Measured bandpass characteristic (S21) in two different network configurations.

band. The measured transmission (S21) characteristic of the filter is shown in figure 1.18 (b). When the MEMS switches are not actuated (short U-shaped elements), the passed band is centred around 6.3 GHz. On the other hand, when the MEMS micro-switches are pulled-in (long U-shaped elements), the passed band is centred around 5.5 GHz, i.e. about 1 GHz below.

The scientific literature is full of several significant contributions exploiting RF-MEMS devices to realize high-performance tunable/switchable RF filters. Differing from other classes of RF-MEMS complex networks, hybridization/integration of MEMS with other technologies has been explored quite broadly. To this regard, several examples are reported concerning reconfigurable filters entirely realized in RF-MEMS technology [87–89]. Besides this option, different ways to obtain integrated RF-MEMS (commercially available) micro-switches with filters realized, for instance, in Printed Circuit Board (PCB) technology, were also investigated, in order to enable the desired tunability [90–93]. Furthermore, another technological solution of particular interest concerns the realization of 3D (evanescent mode) resonant cavities with a high Q-factor, which exploits RF-MEMS tunable elements (mainly varactors) to modify the filter bandpass characteristic [94–97].

RF-MEMS reconfigurable phase shifters

Another class of passive devices that significantly benefits from the exploitation of RF-MEMS technology is that of reconfigurable phase shifters, particularly suited in the driving chain of electronically steerable antennas. An example of an RF-MEMS 5-bit reconfigurable phase shifter in a microstrip configuration is discussed in [98], and its microphotograph is shown in figure 1.19 (a). The device features 5 switchable stages (i.e. 5-bit), in which two paths of different lengths can be selected (per each module) by means of RF-MEMS series ohmic switches. The longer denotes the path the RF signal has to travel across, and the larger denotes the phase shift of the output signal with respect to the input, observable through the S21 parameter. The reconfigurable phase shift of each stage (bit) is added to the others, as the 5 blocks are cascaded, thus leading to 32 possible configurations.

The plot reported in figure 1.19 (b) shows the measured phase shift (S21) of the RF-MEMS network for a few different configurations, in the frequency range from

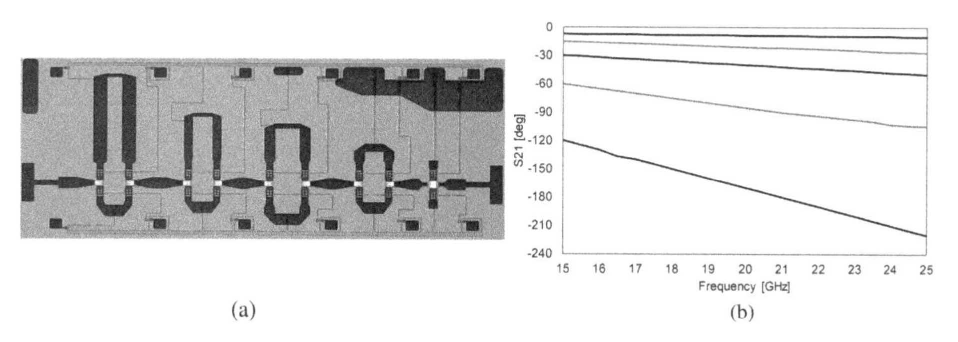

Figure 1.19. (a) Microphotograph of an RF-MEMS 5-bit reconfigurable phase shifter in a microstrip configuration. (b) Input/output phase shift (S21) for different network configurations from 15 GHz to 25 GHz.

15 GHz to 25 GHz. Besides the previously discussed example, the scientific literature reports a wide variety of multi-state RF-MEMS phase shifters. In particular, relevant efforts were devoted to the development of multi-state digital devices [99–101], as well as of continually tunable networks [102–103]. Also interestingly, monolithic solutions in which the RF-MEMS phase shifter is designed and manufactured together with miniaturized reconfigurable antennas within the same technology platform, are reported [104–106].

RF-MEMS impedance matching tuners

The availability of high-performance tunable reactive components, like varactors and inductors, as well as the ease of selection/deselection of fixed high Q-factor capacitances/inductances by means of low-loss ohmic switches, also stimulated the exploitation of RF-MEMS technology for the realization of impedance matching tuners. An example of an impedance matching network entirely realized in RF-MEMS technology is discussed in [107], and the microphotograph of a physical sample is reported in figure 1.20 (a). The device, designed in a CPW configuration, features 8 switching stages, based on cantilever-type RF-MEMS ohmic switches that, depending on their ON/OFF configuration, select different reactive components that load the RF line.

The RF-MEMS network is provided with two banks of reactive components, i.e. Metal-Insulator-Metal (MIM) capacitors, as visible in the upper part of figure 1.20 (a), and in-air suspended inductors, as visible in the lower part of figure 1.20 (a). These reactive elements are designed to be inserted both in series or shunt configuration on the central RF line, depending on which switches are pulled-in and which are left in their rest position. In summary, the RF-MEMS network realizes a double cascaded LC-ladder scheme [85], in which any series or shunt reactive element can be capacitive, inductive, both in parallel, or neither selected, thus enabling 256 different impedance transformations. The Smith chart shown in

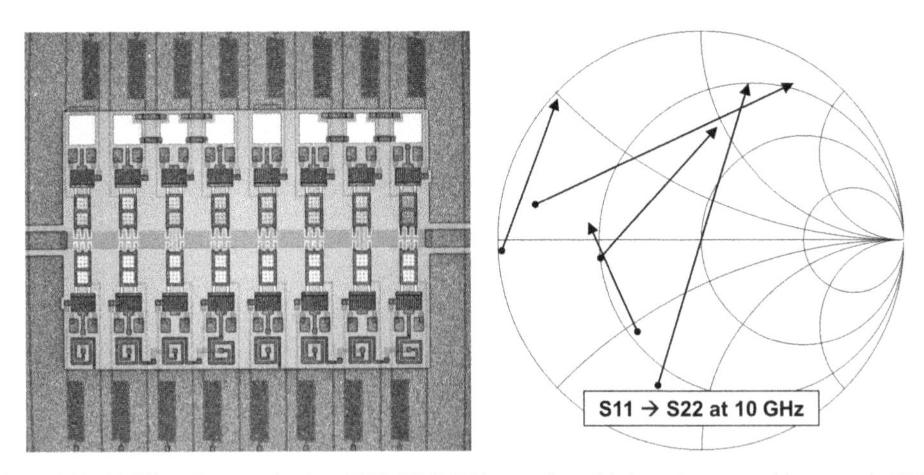

Figure 1.20. (a) Microphotograph of an RF-MEMS 8-bit reconfigurable impedance matching tuner in CPW configuration. (b) Smith chart showing the transformation operated by the device between the input (S11) and output (S22) characteristic impedance in a few configurations, at 10 GHz.

figure 1.20 (b) reports just a few among all functions performed by the impedance tuner. In the plot, each arrow shows how, per each configuration, the input characteristic impedance (S11) is transformed at the output (S22).

Despite not being densely populated as is the case for other classes of devices, the scientific literature discusses several relevant examples of RF-MEMS impedance matching tuners, proving quite extensive coverage of the Smith chart [108–111].

RF-MEMS programmable power attenuators

A category of complex networks that, despite being sporadically studied in the early years of RF-MEMS and recently attracting more attention in the research and industrial scientific community, is that of programmable (step) power attenuators for RF, microwave and millimetre-wave signals. One of the first examples discussed in the literature is reported in [112], while the microphotograph of the fabricated RF-MEMS network is shown in figure 1.21 (a).

The RF-MEMS network features polycrystalline silicon buried resistors of different values, inserted along the RF line (series configuration). Moreover, cantilever-type MEMS ohmic switches, with contact fingers before and after each loading resistor, are placed transversally along the device. When a MEMS switch is actuated (pulled-in), the fingers establish ohmic contact with the underneath metal pads, and a very-low resistance path makes the RF signal flow through the gold switch membrane itself, rather than through the polycrystalline silicon buried resistor. In other words, the loading resistor is shorted, and therefore the attenuation implemented by the whole network is decreased. With particular reference to the device in figure 1.21 (a), the whole attenuation can be stepped according to 4-bit, i.e. 16 different attenuation levels. The plot in figure 1.21 (b) shows the measured attenuation (S21) in a few network configurations, ranging from DC to 30 GHz. Despite the fact that the characterized frequency range is rather wide, all the traces

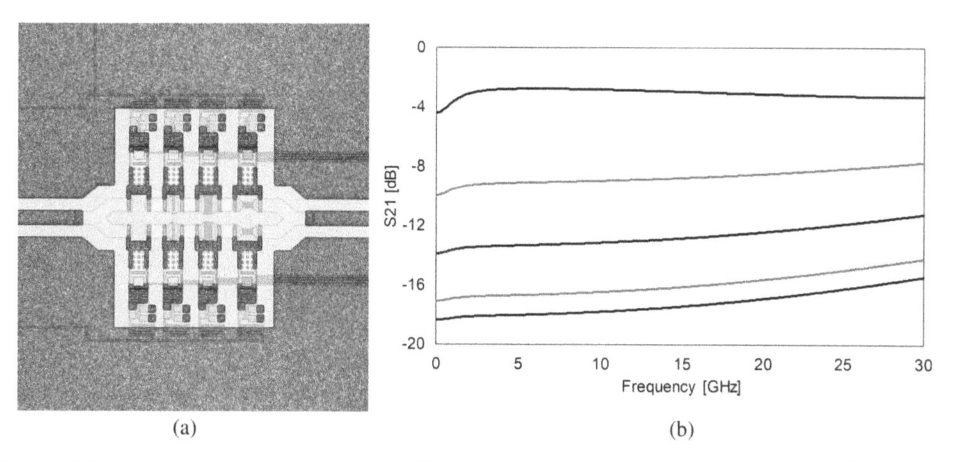

Figure 1.21. (a) Microphotograph of an RF-MEMS 4-bit programmable step attenuator in a CPW configuration. (b) Measured attenuation (S21) realized by the device in a few configurations, from DC to 30 GHz.

exhibit a quite flat characteristic and attenuation levels that can be set from around −3 dB to −20 dB.

Other implementations in RF-MEMS technology of multi-state step attenuators have been recently demonstrated in the literature, in some cases for measured frequency ranges as high as 110 GHz [113–115].

Miscellaneous RF-MEMS

In order to conclude this introductory chapter on RF-MEMS technology, other examples of basic components and complex networks, not covered by the previously reported classes, are going to be briefly listed.

The first category is of mechanical resonators based on MEMS technology. The reasons why they are not always classified as RF-MEMS are multiple. For one thing, often the frequency of operation is not in the RF/microwave/millimetre-wave range, but is lower. In addition, MEMS mechanical resonators are rarely tunable, and they are typically not framed in a waveguide configuration and do not feature switches or varactors. Beside these considerations purely related to a matter of definition, MEMS mechanical resonators are high-performance transducers exploiting a double conversion between physical domains, in order to operate a frequency selection in a very-narrow band. In more detail, a signal with a certain spectrum feeds the resonator, operating electrical to mechanical transduction. The MEMS device resonates at a certain mechanical frequency that is transduced back from a mechanical to electrical domain. The output electrical signal has a very-narrow band with respect to the input one, and the frequency selection is operated by the mechanical characteristics of the MEMS resonator. Such a very-selective filtering function is critical for devices like oscillators, which are meant to generate a reference (RF) frequency with pronounced stability versus time, and a bandwidth as narrow as possible around the carrier. MEMS mechanical resonators with a high Q-factor and remarkable stability against time and temperature have been widely reported in the literature [116–120]. The principle of the forth and back transduction between the electrical and mechanical domain is exploited also in other classes of frequency filtering devices, like the so-called Surface Acoustic Wave (SAW), Bulk Acoustic Wave (BAW) filters and (thin) Film Bulk Acoustic Resonators (FBARs). In this case, the input electrical signal (with a certain spectrum) is transduced into an acoustic wave that travels across a certain material, and then is transduced back into a narrow band electrical signal [121–124].

Another category of RF-MEMS devices is that of electromagnetic resonators and LC-tanks, necessary, for instance, to operate a precise selection of the RF frequency generated by Voltage Controlled Oscillators (VCOs). Differing from mechanical resonators, in this case there is no transduction from the electrical/electromagnetic to mechanical domain to operate selection functionality. The reactive (capacitive and inductive) characteristics of the device determine its specific resonance that, therefore, shapes the output signal. Electromagnetic resonators and LC-tanks significantly benefit from the intrinsic reconfigurability of RF-MEMS, extensively discussed above. For example, it is sufficient to realize part of the capacitive contribution of LC-tanks by means of an RF-MEMS varactor, in order to enable wide tunability of the filtering function operated by the whole device. The scientific literature describes various implementations of electromagnetic resonators and LC-tanks in RF-MEMS technology, both concerning design concepts in planar technology [125, 126], as well as exploiting 3D (evanescent mode) resonant cavities featuring tunable elements [127, 128].

Other RF-MEMS devices of interest in modern RF components and systems are directional couplers/splitters. These passives are meant to couple different signals in order to perform certain mixing functionalities, e.g. of critical importance in RF transmitters/receivers (transceivers), as well as in Automated Test Equipment (ATE) systems, like Vector Network Analysers (VNAs). As it is easy to envisage, the intrinsic tunability of RF-MEMS enables pronounced reconfigurability of couplers/splitters, both concerning selection of different RF signals to be mixed together, as well as with reference to the extent (in terms of dB) according to which such signals must be coupled or split. As for other categories of RF-MEMS, the scientific literature on directional couplers/splitters is populated by several significant contributions, discussing various design concepts and diverse solutions in terms of microfabrication technology platforms [129–135].

Returning to programmable phase shifters, a sub-class of RF-MEMS devices is the so-called True-Time Delay Lines (TTDLs). These reconfigurable networks are, in fact, phase shifters with a quite pronounced linearity of the phase shift over frequency. This characteristic ensures True-Time Delay (TTD), which is a constant (i.e. invariant) delay with respect to frequency. Relevant papers discuss TTDLs in RF-MEMS technology, as well as the achieved remarkable characteristics in terms of reconfigurability [136–140].

Finally, to conclude this chapter, the flexibility of RF-MEMS technology in implementing treatment functionalities of RF/microwave/millimetre-wave signals has unequivocally emerged, in enabling high-performance, pronounced reconfigurability/tunability and wideband operation over frequency. In the following approaches, similar to those discussed in the previous sections, RF designers and engineers now have the possibility to conceive innovative design concepts dedicated to the implementation of RF-MEMS passives with characteristics that transcend those reported up to now. Additionally, blending more functionalities, such as, for example, reconfigurable phase shifting with multi-level step attenuation, within the same RF-MEMS physical device, is also a viable option holding a not-so-hidden potential, especially bearing in mind future 5G applications.

Conclusion

This chapter developed a general discussion on MEMS (MicroElectroMechanical-Systems) and RF-MEMS (MEMS for Radio Frequency passives) technologies. First, the inception of the concept of microsystems was analysed with reference to the evolution of semiconductor technologies, highlighting common features as well as how they can be differentiated. To this regard, a brief discussion was developed around the evolution of (standard) semiconductor technologies, also including the driving trend of Moore's law, that basically described, and is still describing, decades of transistors' technology evolution. It was also remarked that, in parallel to the maturation and consolidation of standard semiconductor technologies, experimentation of microfabrication steps devoted to obtaining microsystems, i.e. micro-devices with mechanical properties, began in the late 1960s. Then, the first examples of actual MEMS devices, reported in the literature in the second half of the 1970s, were also discussed.

The most widely exploited technology process flows for the realization of MEMS, namely, surface and bulk micromachining, were also described, focusing on their fundamental characteristics and on the typical features distinguishing one from the other. Still referring to the intricate relationship between MEMS and standard semiconductor technologies, the concepts of '*More Moore*' and of '*More than Moore*' were debated, putting under the spotlight the opposite trends that micro-systems and semiconductors follow concerning miniaturization and customization.

The concept of RF-MEMS was then introduced, stressing the relative novelty of this kind of microsystem exploitation as compared to sensors and actuators, such as, for example, inertial sensors (accelerometers and gyroscopes) and micro-mirrors.

The four main driving mechanisms to control RF-MEMS devices (electrostatic; electromagnetic; piezoelectric; thermoelectric) were reviewed. Fundamental physical considerations were also included regarding electrostatic actuation of RF-MEMS, it being one of the most diffused, as well as that exploited in the practical examples that will be listed throughout this book.

Subsequently, the most common categories of RF-MEMS devices were introduced, focusing both on simple components, as well as on complex high-order reconfigurable networks. Concerning the former, micro-relays (or switches), variable capacitors (varactors) and inductors were reported. With reference to the latter, instead, several classes of complex RF-MEMS networks were discussed, among which are complex switching units, programmable step attenuators, impedance matching tuners, and so on. For all of the aforementioned categories of RF-MEMS, typical design implementations and experimentally observed performance/characteristics have been shown.

References

[1] Brock D C and Moore G E (ed) 2006 *Understanding Moore's Law: Four Decades of Innovation* 1st edn (Philadelphia, PA: Chemical Heritage Foundation) 122

[2] Converter Passion *ADC Performance Evolution: Low-Voltage Operation—part 2* https://converterpassion.wordpress.com/tag/cmos/ accessed 30 April 2017

[3] Zhang G Q and van Roosmalen A (ed) 2009 *More than Moore—Creating High Value Micro/Nanoelectronics Systems* 1st edn (Berlin: Springer) p 332

[4] Hobstetter J H 1960 Mechanical properties of semiconductors *Properties of Crystalline Solids* 1st edn (West Conshohocken, PA: ASTM) p 40

[5] Beams J W, Freazeale J B and Bart W L 1955 Mechanical strength of thin films of metals *Phys. Rev. Lett.* **100** 1657–61

[6] Neugebauer C A 1960 Tensile properties of thin, evaporated gold films *J. Appl. Phys.* **31** 1096–101

[7] Blakely J M 1964 Mechanical properties of vacuum-deposited gold *J. Appl. Phys.* **35** 1756–9

[8] Bassous E 1978 Fabrication of novel three-dimensional microstructures by the anisotropic etching of (100) and (110) silicon *IEEE Trans. Electron. Dev.* **25** 1178–85

[9] Wen H K, Hynecek J and Boettcher S F 1979 Development of a miniature pressure transducer for biomedical applications *IEEE Trans. Electron. Dev.* **26** 1896–905

[10] Roylance L M and Angell B J 1979 A batch-fabricated silicon accelerometer *IEEE Trans. Electron. Dev.* **26** 1911–17

[11] Chen P, Muller R, Shiosaki T and White R 1979 WP-B6 silicon cantilever beam accelerometer utilizing a PI-FET capacitive transducer *IEEE Trans. Electron. Dev.* **26** 1857

[12] Petersen K E 1979 Micromechanical membrane switches on silicon *IBM J. Res. Dev.* **23** 376–85

[13] Petersen K E 1977 Micromechanical light modulator array fabricated on silicon *Appl. Phys. Lett.* **31** 521

[14] Petersen K E 1982 Silicon as a mechanical material *Proc. IEEE* **70** 420–57

[15] Bustillo J, Howe R and Muller R 1998 Surface micromachining for microelectromechanical systems *Proc. IEEE* **86** 1552–74

[16] Howe R T and Muller R S 1983 Polycrystalline silicon micromechanical beams *J. Electrochem. Soc.* **130** 1420–3

[17] Howe R and Muller R 1986 Resonant-microbridge vapor sensor *IEEE Trans. Electron. Dev.* **33** 499–506

[18] Fan L S, Tai Y C and Muller R S 1987 *Pin Joints, Gears, Springs, Cranks, and other Novel Micromechanical Structures* (Oak Ridge, TN: OSTI)) https://www.osti.gov/scitech/biblio/5974687

[19] Fan L S, Tai Y C and Muller R S 1989 IC-processed electrostatic micromotors *Sensors Actuators* **20** 41–7

[20] Jaeger R C 1988 *Introduction to Microelectronic Fabrication* 1st edn (Boston, MA: Addison-Wesley) p 232

[21] Gualtieri D Interdisciplinarity in Physics *Tikalon Blog* http://tikalon.com/blog/blog.php?article=2012/interdiscipline accessed 3 May 2017

[22] Li H, Li D, Hu Y, Pan P, Li T and Feng J 2016 UV-LIGA microfabrication for high frequency structures of a 220GHz TWT amplifier *Proc. IEEE Int. Vacuum Electron. Conf. (IVEC) (Monterey, CA, April 2016)* pp 1–3

[23] Rusch C, Börner M, Mohr J, Zwick T, Chen Y and De Los Santos H J 2013 Electrical tuning of dielectric resonators with LIGA-MEMS *Proc. Europ. Microw. Integrated Circ. Conf. (Nuremberg, Oct. 2013)* pp 316–9

[24] Johnstone R W and Parmaswaran A 2004 *An Introduction to Surface-Micromachining* 1st edn (Berlin: Springer) p 189

[25] Franssila S 2010 *Introduction to Microfabrication* 2nd edn (New York: John Wiley) 534

[26] Iannacci J, Tschoban C, Reyes J, Maaß U, Huhn M, Ndip I and Pötter H 2016 RF-MEMS for 5G mobile communications: A basic attenuator module demonstrated up to 50 GHz *Proc. IEEE SENSORS (Orlando, FL, Oct.–Nov. 2016)* pp 1–3

[27] Pal P and Sato K (ed) 2017 *Silicon Wet Bulk Micromachining for MEMS* 1st edn (Boca Raton, FL: CRC Press) p 424

[28] Tilli M, Motooka T, Airaksinen V-M, Franssila S, Paulasto-Krockel M and Lindroos V (ed) 2015 *Handbook of Silicon Based MEMS Materials and Technologies* 2nd edn (Norwich, NY: William Andrew) p 826

[29] Iannacci J, Sordo G, Serra E and Schmid U 2016 The MEMS four-leaf clover wideband vibration energy harvesting device: design concept and experimental verification *Microsyst. Technol.* **22** 1865–81
[30] Sherman S J, Tsang W K, Core T A and Quinn D E 1992 A low cost monolithic accelerometer *Proc. of VLSI Circ. Symp. (Seattle, WA, June 1992)* pp 34–5
[31] Iannacci J 2015 Reliability of MEMS: A perspective on failure mechanisms, improvement solutions and best practices at development level *Displays* **37** 62–71
[32] Sensors Online *The Growing Presence of MEMS and MST in Automotive Applications* http://archives.sensorsmag.com/articles/0999/89/main.shtml accessed 4 May 2017
[33] National Inventors Hall of Fame *Larry Hornbeck, Digital Micromirror Device* http://www.invent.org/honor/inductees/inductee-detail/?IID=397 accessed 4 May 2017
[34] Douglass M R 1998 Lifetime estimates and unique failure mechanisms of the Digital Micromirror Device (DMD) *Proc. of IEEE Int. Reliab. Phys. Symp. (Reno, NV, April 1998)* pp 9–16
[35] SolidState Technology *The Inertial MEMS Device Market Keeps on Growing. What's Next?* http://electroiq.com/blog/2015/11/the-inertial-mems-device-market-keeps-on-growing-whats-next/ accessed 4 May 2017
[36] Econocom *How the Internet of Things is Revolutionising Business Models* https://blog.econocom.com/en/blog/how-the-internet-of-things-is-revolutionising-business-models/ accessed 4 May 2017
[37] McGrath W R, Walker C, Yap M and Tai Y C 1993 Silicon micromachined waveguides for millimeter-wave and submillimeter-wave frequencies *IEEE Microw. Guid. Wave Lett* **3** 61–3
[38] Katehi L P B, Rebeiz G M, Weller T M, Drayton R F, Cheng H J and Whitaker J F 1993 Micromachined circuits for millimeter- and sub-millimeter-wave applications *IEEE Antennas Propag. Mag.* **35** 9–17
[39] Weller T M and Katehi L P B 1995 Compact stubs for micromachined coplanar waveguide *Proc. 25th Europ. Microw. Conf. (Bologna, Sept. 1995)* pp 589–93
[40] Mahmoud S F 1991 *Electromagnetic Waveguides: Theory and Applications* 1st edn (London: Peter Peregrinus) p 240
[41] Iannacci J 2013 RF passive components for wireless applications *Handbook of MEMS for Wireless and Mobile Applications* 1st edn (Cambridge: Woodhead Publishing) pp 100–35
[42] Pozar D M 2005 *Microwave Engineering* 2nd edn (New York: John Wiley) p 716
[43] Wadell B 1991 *Transmission Line Design Handbook* 1st edn (Norwood: Artech House)
[44] Yang S, Hu Z, Buchanan N B, Fusco V F, Stewart J A C, Wu Y, Armstrong B M, Armstrong G A and Gamble H S 1998 Characteristics of trenched coplanar waveguide for high-resistivity Si MMIC applications *IEEE Trans. Microw. Theory Tech* **46** 623–31
[45] Farrington N E S and Iezekiel S 2011 Design and simulation of membrane supported transmission lines for interconnects in a MM-wave multichip module *Prog. Electromag. Res.* B **27** 165–86
[46] Shi Y, Lai Z, Xin P, Shao L and Zhu Z 2001 Design and fabrication of micromachined microwave transmission lines *Proc. SPIE* **4557** 1–8
[47] Goldsmith C L, Yao Zhimin, Eshelman S and Denniston D 1998 Performance of low-loss RF MEMS capacitive switches *IEEE Microw. Guid. Wave Lett.* **8** 269–71
[48] Feng Z, Zhang W, Su B, Harsh K F, Gupta K C, Bright V and Lee Y C 1999 Design and modeling of RF MEMS tunable capacitors using electro-thermal actuators *Proc. IEEE MTT-S Int. Microw. Symp. (Anaheim, CA, June 1999)* pp 1507–10

[49] Katehi L P B, Rebeiz G M and Nguyen C T-C 1998 MEMS and Si-micromachined components for low-power, high-frequency communications systems *Proc. IEEE MTT-S Int. Microw. Symp. (Baltimore, MD, June 1998)* pp 331–3
[50] Malczewski A, Eshelman S, Pillans B, Ehmke J and Goldsmith C L 1999 X-band RF MEMS phase shifters for phased array applications *IEEE Microw. Guid. Wave Lett.* **9** 517–9
[51] Liu C 2011 *Foundations of MEMS* 2nd edn (London: Pearson Education) p 560
[52] Lee H S, Leung C H, Shi J and Chan S C 2004 Micro-electro-mechanical relays-design concepts and process demonstrations *Proc. 50th IEEE Holm Conf. on Electrical Contacts and the 22nd Int. Conf. on Electrical Contacts (Seattle, WA, Sept. 2004)* pp 242–7
[53] Cho I-J, Song T, Baek S-H and Yoon E 2005 A low-voltage and low-power RF MEMS series and shunt switches actuated by combination of electromagnetic and electrostatic forces *IEEE Trans. Microw. Theory Tech* **53** 2450–7
[54] Safari A and Akdoğan E K (ed) 2008 *Piezoelectric and Acoustic Materials for Transducer Applications* 1st edn (New York: Springer) p 482
[55] Kawakubo T, Nagano T, Nishigaki M, Abe K and Itaya K 2005 Piezoelectric RF MEMS tunable capacitor with 3V operation using CMOS compatible materials and process *Proc. IEEE Int. Electron Dev. Meet. IEDM (Washington, DC, Dec. 2005)* pp 294–7
[56] Daneshmand M, Fouladi S, Mansour R R, Lisi M and Stajcer T 2009 Thermally-actuated latching RF MEMS switch *Proc. IEEE Int. Microw. Symp. MTT-S Digest (Boston, MA, June 2009)* pp 1217–20
[57] Iannacci J, Faes A, Repchankova A, Tazzoli A and Meneghesso G 2011 An active heat-based restoring mechanism for improving the reliability of RF-MEMS switches *Microelectron. Reliab.* **51** 1869–73
[58] Iannacci J, Repchankova A, Faes A, Tazzoli A, Meneghesso G and Dalla Betta G F 2010 Enhancement of RF-MEMS switch reliability through an active anti-stiction heat-based mechanism *Microelectron. Reliab.* **50** 1599–603
[59] Senturia S D 2001 *Microsystem Design* 1st edn (New York: Springer) p 689
[60] Giacomozzi F, Mulloni V, Colpo S, Iannacci J, Margesin B and Faes 2011 A flexible fabrication process for RF MEMS devices *Rom. J. Inf. Sci. Technol. (ROMJIST)* **14** 259–68
[61] Iannacci J, Resta G, Farinelli P and Sorrentino R 2012 RF-MEMS components and networks for high-performance reconfigurable telecommunication and wireless systems *Trans. Tech. Pub. Adv. Sci. Tech* **81** 65–74
[62] Iannacci J 2013 *Practical guide to RF-MEMS* 1st edn (Weinheim: Wiley-VCH) p 372
[63] Carty E, Fitzgerald P, McDaid P, Stenson B and Goggin R 2016 Development of a DC to K-band ultra long on-life RF MEMS switch with integrated driver circuitry *Proc. 11th Europ. Microw, Integrated Circ. Conf. EuMIC (London, Oct. 2016)* pp 444–7
[64] Seki T, Yamamoto J, Murakami A, Yoshitake N, Hinuma K-i, Fujiwara T, Sano K, Matsushita T, Sato F and Oba M 2013 An RF MEMS switch for 4G Front-Ends *Proc. IEEE MTT-S Int. Microw. Symp. IMS (Seattle, WA, June 2013)* pp 1–3
[65] Chu C, Liao X and Yan H 2017 Ka-band RF MEMS capacitive switch with low loss, high isolation, long-term reliability and high power handling based on GaAs MMIC technology *IET Microw. Antennas Propag.* **11** 942–8

[66] Ma L Y, Soin N and Nordin A N 2016 A novel design of low-voltage low-loss K-band RF-MEMS capacitive switch *Proc. Symp. on Design, Test, Integr. and Packaging of MEMS/MOEMS DTIP (Budapest, May–June 2016)* pp 1–5

[67] Iannacci J, Gaddi R and Gnudi A 2007 Non-linear electromechanical RF model of a MEMS varactor based on veriloga© and lumped-element parasitic network *Proc. Europ. Microw. Integ. Circ. Conf. (Munich, Oct. 2007)* pp 544–7

[68] Elshurafa A M, Ho P H and Salama K N 2012 Low voltage RF MEMS variable capacitor with linear C–V response *IET Electron. Lett.* **48** 392–3

[69] Nishiyama M, Konishi H, Suzuki J, Tezuka Y, Suzuki Y and Suzuki K 2007 Extremely high capacitance ratio (C/R) RF MEMS variable capacitor with chameleon actuators *Proc. Int. Solid-State Sensors, Actuators and Microsyst. Conf. TRANSDUCERS (Lyon, June 2007)* pp 631–4

[70] Cazzorla A, Sorrentino R and Farinelli P 2015 Double-actuation extended tuning range RF MEMS Varactor *Proc. Europ. Microw. Conf. EuMC (Paris, Oct. 2015)* pp 937–40

[71] Chen K, Kovacs A and Peroulis D 2010 Anti-biased RF MEMS varactor topology for 20–25 dB linearity enhancement *Proc. IEEE MTT-S Int. Microw. Symp. (Anaheim, CA, May 2010)* pp 1142–5

[72] Ikehashi T, Miyazaki T, Yamazaki H, Suzuki A, Ogawa E, Miyano S, Saito T, Ohguro T, Miyagi T, Sugizaki Y, Otsuka N, Shibata H and Toyoshima Y 2008 An RF MEMS variable capacitor with intelligent bipolar actuation *Proc. IEEE Int. Solid-State Circ. Conf. (San Francisco, CA, Feb. 2008)* pp 582–637

[73] Mizuochi Y, Amakawa S, Ishihara N and Masu K 2009 Study of air-suspended RF MEMS inductor configurations for realizing large inductance variations *Proc. Argentine School of Micro-Nanoelectronics, Tech. and Applications (San Carlos de Bariloche, Oct. 2009)* pp 50–5

[74] Chang S and Sivoththaman S 2006 A tunable RF MEMS inductor on silicon incorporating an amorphous silicon bimorph in a low-temperature process *IEEE Electron Dev. Lett.* **27** 905–7

[75] Zine-El-Abidine I, Okoniewski M and McRory J G 2004 A tunable RF MEMS inductor *Proc. Int. Conf. on MEMS, NANO and Smart Syst. ICMENS (Banff, Aug 2004)* pp 636–8

[76] Banitorfian F, Eshghabadi F, Manaf A A, Pons P, Noh N M, Mustaffa M T and Sidek O 2013 A novel tunable water-based RF MEMS solenoid inductor *Proc. IEEE Regional Symp. on Micro and Nanoelectron. RSM (Langkawi, Sept. 2013)* pp 58–61

[77] Hikmat O F and Mohamed Ali M S 2017 RF MEMS inductors and their applications—A review *IEEE J. Microelectromech. Syst.* **26** 17–44

[78] Casini F, Farinelli P, Mannocchi G, DiNardo S, Margesin B, De Angelis G, Marcelli R, Vendier O and Vietzorreck L 2010 High performance RF-MEMS SP4T switches in CPW technology for space applications *Proc. Europ. Microw. Conf. (Paris, Sept. 2010)* pp 89–92

[79] Yang H-H, Yahiaoui A, Zareie H, Blondy P and Rebeiz G M 2014 A compact high-isolation DC-50 GHz SP4T RF MEMS switch *Proc. IEEE MTT-S Int. Microw. Symp. IMS (Tampa, FL, June 2014)* pp 1–4

[80] Zareie H and Rebeiz G M 2014 Compact high-power SPST and SP4T RF MEMS metal-contact switches *IEEE Trans. Microw. Theory Tech.* **62** 297–305

[81] Chan K Y, Ramer R and Mansour R R 2012 Novel miniaturized RF MEMS staircase switch matrix *IEEE Microw. Wirel. Comp. Lett.* **22** 117–9

[82] Daneshmand M and Mansour R R 2011 RF MEMS Satellite Switch Matrices *IEEE Microw. Mag.* **12** 92–109

[83] Fomani A A and Mansour R R 2009 Monolithically integrated multiport RF MEMS switch matrices *IEEE Trans. Microw. Theory Tech.* **57** 3434–41

[84] Fomani A A and Mansour R R 2009 Miniature RF MEMS switch matrices *Proc. IEEE MTT-S Int. Microw. Symp. IMS (Boston, MA, June 2009)* pp 1221–4

[85] Besser L and Gilmore R 2003 *Practical RF Circuit Design for Modern Wireless Systems: Passive circuits and systems* 1st edn (Norwood, MA: Artech House) p 576

[86] Ocera A, Farinelli P, Mezzanotte P, Sorrentino R, Margesin B and Giacomozzi F 2006 A novel MEMS-tunable hairpin line filter on silicon substrate *Proc. Europ. Microw. Conf. (Manchester, Sept. 2006)* pp 803–6

[87] Reines I, Park S J and Rebeiz G M 2010 Compact low-loss tunable X-band bandstop filter with miniature RF-MEMS switches *IEEE Trans. Microw. Theory Tech.* **58** 1887–95

[88] Hsu H H, Margomenos A D and Peroulis D 2011 A monolithic RF-MEMS filter with continuously-tunable center-frequency and bandwidth *Proc. IEEE Top. Meet. on Silicon Monolithic Integr. Circ. in RF Syst. (Phoenix, AZ, Jan. 2011)* pp 169–72

[89] Shah U, Sterner M and Oberhammer J 2011 Basic concepts of moving-sidewall tuneable capacitors for RF MEMS reconfigurable filters *Proc. Europ. Microw. Integ. Circ. Conf. (Manchester, Oct. 2011)* pp 526–9

[90] Shojaei-Asanjan D and Mansour R R 2017 The sky's the limit: A switchable RF-MEMS filter design for wireless avionics intracommunication *IEEE Microw. Mag.* **18** 100–6

[91] Wang H, Anand A and Liu X 2017 A miniature 800–1100-MHz tunable filter with high-Q ceramic coaxial resonators and commercial RF-MEMS tunable digital capacitors *Proc. IEEE Wireless and Microw. Tech. Conf. WAMICON (Cocoa Beach, FL, April 2017)* pp 1–3

[92] Kumar N and Singh Y K 2017 RF-MEMS-based bandpass-to-bandstop switchable single- and dual-band filters with variable FBW and reconfigurable selectivity *IEEE Trans. Microw. Theory Tech.* pp 1–14

[93] Hickle M D, Li J, Psychogiou D and Peroulis D 2016 A high-performance pathway: a 0.95/2.45-GHZ switched-frequency bandpass filter using commercially available RF MEMS tuning elements *IEEE Microw. Mag.* **17** 34–41

[94] Stefanini R, Chatras M, Pothier A, Guines C and Blondy P 2013 High-Q 3D tunable RF MEMS filter with a constant fractional bandwidth *Proc. Europ. Microw. Integr. Circ. Conf. (Nuremberg, Oct. 2013)* pp 312–5

[95] Chan K Y, Ramer R and Mansour R R 2017 A switchable iris bandpass filter using RF MEMS switchable planar resonators *IEEE Microw. Wirel. Comp. Lett.* **27** 34–6

[96] Park S J, Reines I, Patel C and Rebeiz G M 2010 High-Q RF-MEMS 4–6-GHz tunable evanescent-mode cavity filter *IEEE Trans. Microw. Theory Tech.* **58** 381–9

[97] Schulte B, Ziegler V, Schoenlinner B, Prechtel U and Schumacher H 2011 RF-MEMS tunable evanescent mode cavity filter in LTCC technology at Ku-band *Proc. of Europ. Microw. Integr. Circ. Conf. (Manchester, Oct. 2011)* pp 514–7

[98] Bastioli S, Di Maggio F, Farinelli P, Giacomozzi F, Margesin B, Ocera A, Pomona I, Russo M and Sorrentino R 2008 Design manufacturing and packaging of a 5-bit K-band MEMS phase shifter *Proc. Europ. Microw. Integr. Circ. Conf. (Amsterdam, Oct. 2008)* pp 338–41

[99] Ramli N A and Arslan T 2017 Design and simulation of a 2-bit distributed S-band MEMS phase shifter *Proc. Int. Conf. Thermal, Mech. and Multi-Physics Simul. and Exp. in Microelectron.and Microsyst. EuroSimE (Dresden, April 2017)* pp 1–5

[100] Dey S and Koul S K 2014 10–35-GHz frequency reconfigurable RF MEMS 5-bit DMTL phase shifter uses push-pull actuation based toggle mechanism *Proc. IEEE Int. Microw. and RF Conf. IMaRC (Bangalore, Dec. 2014)* pp 21–4

[101] Bakri-Kassem M, Mansour R R and Safavi-Naeini S 2014 A novel latching RF MEMS phase shifter *Proc. Europ. Microw. Integr. Circ. Conf. (Rome, Oct. 2014)* pp 668–71

[102] McFeetors G and Okoniewski M 2004 Analog tunable microwave phase shifters using RF MEMS *Proc. Int. Symp. on Antenna Tech. and Appl. Electromagnetics and URSI Conf. (Ottawa, July 2004)* pp 1–4

[103] Unlu M, Demir S and Akin T 2013 A 15–40-GHz frequency reconfigurable RF MEMS phase shifter *IEEE Trans. Microw. Theory Tech.* **61** 2865–77

[104] Abumunshar A J, Nahar N K, Hyman D and Sertel K 2017 18–40GHz low-profile phased array with integrated MEMS phase shifters *Proc. Europ. Conf. on Antennas and Propagation EUCAP (Paris, March 2017)* pp 2800–1

[105] Yeo W G, Nahar N K and Sertel K 2013 Phased array antenna with integrated MEMS phase shifters for Ka-band SATCOM *Proc. IEEE Antennas and Propagation Soc. Int. Symp. APSURSI (Orlando, FL, July 2013)* pp 105–6

[106] Das A, Puri M and Sengar J S 2013 A novel monolithic integrated phased array antenna using 4-bit distributed MEMS phase shifter and triangular patch antenna *Proc. Int. Conf. on Adv. in Comput., Commun. and Informatics ICACCI (Mysore, Aug. 2013)* pp 913–8

[107] Iannacci J, Masotti D, Kuenzig T and Niessner M 2011 A reconfigurable impedance matching network entirely manufactured in RF-MEMS technology *Proc. SPIE* **8066** 1–12

[108] Festo A E, Folgero K, Ullaland K and Gjertsen K M 2009 A six bit, 6–18 GHz, RF-MEMS impedance tuner for 50 Ω systems *Proc. Europ. Microw. Conf. EuMC (Rome, Sept.–Oct. 2009)* pp 1132–5

[109] Vaha-Heikkila T, Varis J, Tuovinen J and Rebeiz G M 2004 A V-band single-stub RF MEMS impedance tuner *Proc. Europ. Microw. Conf. (Amsterdam, Oct. 2004)* pp 1301–4

[110] Fouladi S, Domingue F and Mansour R 2012 CMOS-MEMS tuning and impedance matching circuits for reconfigurable RF front-ends *Proc. IEEE/MTT-S Int. Microw. Symp. (Montreal, June 2012)* pp 1–3

[111] Larcher L, Brama R, Ganzerli G, Iannacci J, Margesin B, Bedani B and Gnudi A 2009 A MEMS reconfigurable quad-band class-E power amplifier for GSM standard *Proc. IEEE Int. Conf. on Micro Electro Mech. Syst. (Sorrento, Jan. 2009)* pp 864–7

[112] Iannacci J, Faes A, Mastri F, Masotti D and Rizzoli V 2010 A MEMS-based wide-band multi-state power attenuator for radio frequency and microwave applications *Proc. TechConnect World, NSTI Nanotech (Anaheim, CA, June 2010)* pp 328–31

[113] Sun J and Li Z 2016 A broadband 3-bit MEMS digital attenuator *Proc. Int. Conf. on Electron. Design ICED (Phuket, Aug 2016)* pp 442–5

[114] Iannacci J, Huhn M, Tschoban C and Pötter H 2016 RF-MEMS technology for future mobile and high-frequency applications: Reconfigurable 8-bit power attenuator tested up to 110 GHz *IEEE Electron Dev. Lett.* **37** 1646–9

[115] Iannacci J, Huhn M, Tschoban C and Pötter H 2016 RF-MEMS technology for 5G: Series and shunt attenuator modules demonstrated up to 110 GHz *IEEE Electron Dev. Lett.* **37** 1336–9

[116] Courcimault C G and Allen M G 2005 High-Q mechanical tuning of MEMS resonators using a metal deposition-annealing technique *Proc. Int. Conf. on Solid-State Sensors, Actuators and Microsyst. TRANSDUCERS (Seoul, June 2005)* pp 875–8

[117] Pillai G, Tan W S, Chen C C and Li S S 2016 Modeling of zero TCF and maximum bandwidth orientation for lithium tantalate RF MEMS resonators *Proc. IEEE Ann. Int. Conf. on Nano/Micro Engineered and Molecular Syst. NEMS (Sendai, April 2016)* pp 450–4

[118] Pacheco S, Zurcher P, Young S, Weston D and Dauksher W 2004 RF MEMS resonator for CMOS back-end-of-line integration *Proc. Top. Meet. on Silicon Monolithic Integ. Circ. in RF Syst. (Atlanta, GA, Sept. 2004)* pp 203–6

[119] Cassella C, Chen G, Qian Z, Hummel G and Rinaldi M 2017 RF passive components based on aluminum nitride cross-sectional Lamé-mode MEMS resonators *IEEE Trans. Electron Dev.* **64** 237–43

[120] Yuan Q, Luo W, Zhao H, Peng B, Yang J and Yang F 2015 Frequency stability of RF-MEMS disk resonators *IEEE Trans. Electron Dev.* **62** 1603–8

[121] Hashimoto K-y 2009 *RF Bulk Acoustic Wave Filters for Communications* 1st edn (Norwood: Artech House) p 275

[122] Enz C C and Kaiser A (ed) 2013 *MEMS-based Circuits and Systems for Wireless Communication* 1st edn (New York: Springer) p 332

[123] Aigner R 2008 SAW and BAW technologies for RF filter applications: A review of the relative strengths and weaknesses *Proc. IEEE Ultrasonics Symp. (Beijing, Nov. 2008)* pp 582–9

[124] Koohi M Z, Lee S and Mortazawi A 2016 Design of BST-on-Si composite FBARs for switchable BAW filter application *Proc. Europ. Microw. Conf. EuMC (London, Oct. 2016)* pp 1003–6

[125] Gaddi R, Gnudi A, Franchi E, Guermandi D, Tortori P, Margesin B and Giacomozzi F 2005 Reconfigurable MEMS-enabled LC-tank for multi-band CMOS oscillator *Proc. IEEE MTT-S Int. Microw. Symp. IMS (Long Beach, CA, June 2005)* pp 1–4

[126] Cazzorla A, Sorrentino R and Farinelli P 2015 MEMS based LC tank with extended tuning range for multi-band applications *Proc. IEEE Mediterranean Microw. Symp. MMS (Lecce, Nov.–Dec. 2015)* pp 1–4

[127] Small J, Arif M S, Fruehling A and Peroulis D 2013 A tunable miniaturized RF MEMS resonator with simultaneous high Q (500–735) and fast response speed (<10—60 *μs) IEEE J. Microelectromech. Syst.* **22** 395–405

[128] Irshad W and Peroulis D 2011 A 12–18 GHz electrostatically tunable liquid metal RF MEMS resonator with quality factor of 1400–1840 *Proc. IEEE MTT-S Int. Microw. Symp. IMS (Baltimore, MD, June 2011)* pp 1–4

[129] Psychogiou D, Yang Z and Peroulis D 2012 RF-MEMS enabled power divider with arbitrary power division ratio *Proc. Europ. Microw. Conf. (Amsterdam, Oct. 2012)* pp 53–6

[130] Ocera A, Farinelli P, Cherubini F, Mezzanotte P, Sorrentino R, Margesin B and Giacomozzi F 2007 A MEMS-reconfigurable power divider on high resistivity silicon substrate *Proc. IEEE/MTT-S Int. Microw. Symp. IMS (Honolulu, HI, June 2007)* pp 501–4

[131] Ocera A, Gatti R V, Mezzanotte P, Farinelli P and Sorrentino R 2005 A MEMS programmable power divider/combiner for reconfigurable antenna systems *Proc. Europ. Microw. Conf. (Paris, Oct. 2005)* pp 1–4

[132] Marcaccioli L, Farinelli P, Tentzeris M M, Papapolymerou J and Sorrentino R 2008 Design of a broadband MEMS-based reconfigurable coupler in Ku-band *Proc. Europ. Microw. Conf. (Amsterdam, Oct. 2008)* pp 595–8

[133] Gurbuz O D and Rebeiz G M 2015 A 1.6–2.3-GHz RF MEMS reconfigurable quadrature coupler and its application to a 360° reflective-type phase shifter *IEEE Trans. Microw. Theory Tech.* **63** 414–21

[134] Ocera A, Farinelli P, Mezzanotte P, Sorrentino R, Margesin B and Giacomozzi F 2007 Novel RF-MEMS widely-reconfigurable directional coupler *Proc. Europ. Microw. Conf. (Munich, Oct. 2007)* pp 122–5

[135] Shah U, Sterner M and Oberhammer J 2013 High-directivity MEMS-tunable directional couplers for 10–18-GHz broadband applications *IEEE Trans. Microw. Theory Tech.* **61** 3236–46

[136] De Angelis G, Lucibello A, Marcelli R, Catoni S, Lanciano A, Buttiglione R, Dispenza M, Giacomozzi F, Margesin B, Maglione A, Erspan M and Combi C 2008 Packaged single pole double thru (SPDT) and true time delay lines (TTDL) based on RF MEMS switches *Proc. Int. Semicond. Conf. CAS (Sinaia, Oct. 2008)* pp 227–30

[137] Fu D, Bey Y A, Domier C, Luhmann N C and Liu X 2015 A Q-band RF-MEMS tapered true time delay line for fusion plasma diagnostics systems *Proc. IEEE MTT-S Int. Microw. Symp. IMS (Phoenix, AZ, May 2015)* pp 1–3

[138] Van Caekenberghe K and Vaha-Heikkila T 2008 An analog RF MEMS slotline true-time-delay phase shifter *IEEE Trans. Microw. Theory Tech.* **56** 2151–9

[139] Hacker J B, Mihailovich R E, Kim M and DeNatale J F 2003 A Ka-band 3-bit RF MEMS true-time-delay network *IEEE Trans. Microw. Theory Tech.* **51** 305–8

[140] Kim M, Hacker J B, Mihailovich R E and DeNatale J F 2001 A DC-to-40 GHz four-bit RF MEMS true-time delay network *IEEE Microw. Wirel. Comp. Lett.* **11** 56–8

Chapter 2

Expectations versus actual market of RF-MEMS: Analysis and explanation of a repeatedly fluctuating scenario

2.1 General considerations of the market in the technology sector

This chapter will focus on the analysis of the RF-MEMS potentialities in terms of market penetration, by reviewing the expectations and forecasts taking turns in the last 15 years, as well as by interpreting the reasons for them not being fully achieved. Nonetheless, in order for such a detailed study to be effective, it is necessary to first introduce a couple of basic concepts belonging to marketing and management science.

The first notion is the one of the hype curve, which is a graphic tool elaborated by Gartner Inc., an information technology research and advisory company. Such a curve depicts the typical cycle of acceptance related to a new technology, from its first development to consolidated employment in market applications [1]. The usual trend of the hype curve is shown in figure 2.1.

When a novel technology starts to be intensively investigated at the research level and remarkable performance/characteristics are demonstrated in lab-based scenarios (a—*Technology trigger*), its visibility increases and, in parallel to it, expectations of market placement and outcomes become higher (b—*Peak of inflated expectations*). Subsequently, when the efforts of research start approaching real application cases and scenarios, aspects of the new technology that were not critical for demonstration purposes, but that, on the other hand, are crucial from the market perspective, begin to emerge. Among them, the most important ones are typically limited in number, and very often the same, like integration/compatibility with other (already existing) technologies, reliability and manufacturing/production costs. The aforementioned context causes inflated expectations to drop (c—*Trough of disillusionment*). At the same time, such a scenario provides the scientific and industrial community with proper motivation to address and overcome limitations of the new

doi:10.1088/978-0-7503-1545-6ch2

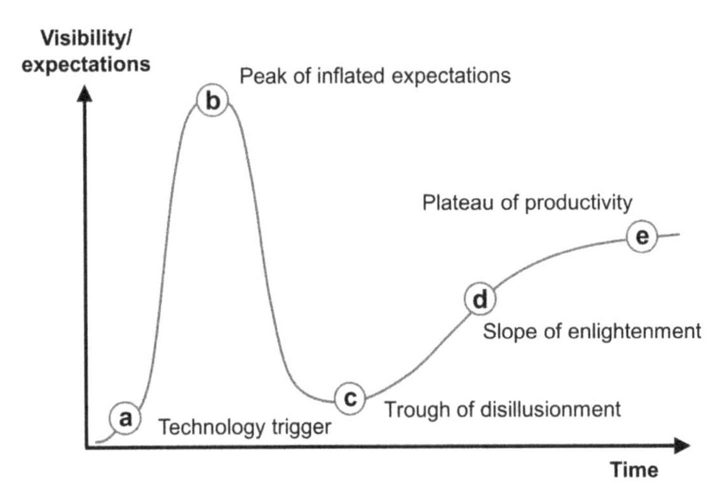

Figure 2.1. Typical hype curve behaviour, describing the evolution of a novel technology against time, since its first investigation (in the basic research scenario), to the maturation and employment in market and commercial applications.

technology against real application scenarios (d—*Slope of enlightenment*). Finally, when the technology is mature enough, it starts to be steadily employed and available on the market, leading to the final stage of the hype curve (e—*Plateau of productivity*). The whole hype cycle exhibits a different duration depending on several factors, such as, for instance, technology complexity, market acceptance, market needs as well as readiness, and so on. In any case, the time unit of the horizontal axis in figure 2.1 is typically in the order of a magnitude of years.

Still referring to the ways in which new technologies penetrate the market, if, on the one hand, such paths can be multiple and quite diverse, in essence they fall within two main scenarios, labelled *technology push* and *market driven* [2].

In the technology push scenario, a novel solution is introduced into the market neither because of a factual need to be addressed, nor of a certain issue to be overcome. In such a context, the added value related to the employment of a certain technology or component can be the extension of system functionality or, more trivially, the perception of distinctiveness it provides to the final user, as happens with the concept of brand. An example of the technology push approach can be found when cameras were integrated in the previous generations of mobile phones. Around fifteen years ago, probably no-one, among the millions of handset users, was expecting to take pictures and make videos with their cellular phones. More generally, it is unlikely that anybody was even thinking to use handsets to do something other than placing calls and typing short (purely textual) messages. Despite not responding to an actual market need, cameras started to be integrated in mobile handsets, and people started to become accustomed to the possibility of accessing and visualizing multimedia content through their portable phones. In other words, still referring to the aforementioned example, the technology push approach does not respond to a market need, but, conversely, aims at its generation.

On the other bank of the river stands the market pull/driven scenario. In this case, the introduction of an innovative technology, or the improvement of an existing one, is driven by specific market needs, such as issues to be overcome or more stringent specifications to be met. Despite looking odd at first glance, a valid example of the market pull scenario is still the one of integrated cameras in mobile handsets. As a matter of fact, after stepping across the user acceptance phase, i.e. the time interval from the introduction of a novel technology to its recognition as being valuable and necessary by most users, it is probable that nobody is currently thinking of using a mobile handset just for calling and text messaging. In other words, pushing cameras into mobile phones contributed, in a matter of a few years, to create a strong market need driving innovation in smartphone technologies for vision, thus aiming at a better quality of images and videos, smaller and more integrated devices, reduced power consumption, and, last but not least, lower prices.

Now that fundamental concepts concerning the market acceptance of novel technologies and solutions have been introduced, the evolution of expected versus real market of RF-MEMS is going to be analysed.

2.2 RF-MEMS on the market: vision of the early days

Since the early discussions in scientific literature, comprehensively reported before in this work, RF-MEMS passive components and networks exhibited remarkable performance and characteristics, definitely boosted if compared against those typical of their counterparts realized in standard (semiconductor) technologies. This aspect provided an enormous stimulus to scientists and engineers in proliferating visions around future (at that time) exploitations of RF-MEMS technology in radio and wireless applications and systems. Several valuable works and ideas were published by distinguished researchers, both in the academic and private sector. Among the former, Clark T-C Nguyen provided a quite extensive contribution that is going to be referenced here as an explanatory example.

In the initial phase, the work developed by Nguyen pioneered the field of MEMS mechanical resonators with a very high quality factor (Q-factor), to be employed in highly selective RF oscillators [3–4]. In parallel to such a development at the device level, possible ways RF-MEMS technology might have exerted an impact on radio transmitters/receivers (transceivers) started to be analysed. To this purpose, one of the most diffused transceiver architectures, namely the super-heterodyne transmitter/receiver [5], was taken as a reference. Such an architecture operates frequency up conversion (transmitter end) and down conversion (receiver end) of the baseband signal, to an Intermediate Frequency (IF) band lower than the RF band exploited for signal broadcasting. The IF eases the treatment of the signal to be transmitted or received (e.g. filtering, amplification, modulation/demodulation, and so on). According to the vision of Nguyen, the impact of RF-MEMS technology was expected to be twofold, with a limited impact on transceiver architectures at first, and with a partial redesign of them in a more advanced phase.

In the first stage, RF-MEMS basic passive components would have replaced traditional devices in several parts of the super-heterodyne transceiver, indeed improving its performance and characteristics [4–6]. The block diagram in figure 2.2 shows the typical configuration of a super-heterodyne receiver, as reported and discussed by Nguyen.

All the coloured (and check-marked) diagram sub-blocks could have been replaced by implementations of passive components in RF-MEMS technology. In particular, looking at the receiver from the antenna to the output of the demodulated received signal—that is both In-phase (I) and Quadrature (Q)—a MEMS switching unit could be used to select the proper antenna. Moreover, MEMS varactors and inductors are suitable to realize high Q-factor filters (both RF and IF) and reconfigurable LC-tanks to tune the oscillation frequency of the Voltage Controlled Oscillator (VCO). Finally, MEMS resonators could replace the typical quartz-based devices within oscillators.

As widely discussed prior to this point, besides the realization of RF-MEMS basic passives, MEMS technology is suitable for implementing more complex elements as well. To this regard, high-order switching matrices, reconfigurable phase shifters, impedance matching tuners, programmable step attenuators, and so on, have already been discussed.

In the second stage forecasted by Nguyen, the availability of such networks would have led to rethinking the architecture of a transceiver, rather than just replacing some of its elemental components with RF-MEMS counterparts. Sticking to the same example as before, the super-heterodyne receiver architecture could have been rearranged and simplified on the basis of RF-MEMS complex networks, as discussed in [6–9] and shown in figure 2.3.

In this altered scheme, a multi-channel selector (i.e. a bank of MEMS switches) with several filtering functions would simplify the hardware complexity of the whole platform. The typical low-loss of MEMS-based devices and networks compared to standard technologies would reduce the number of Power Amplifiers (PAs) needed to regenerate the signal. Moreover, the high-reconfigurability of MEMS could be exploited to realize a widely tuneable oscillator, thus extending the range of possible

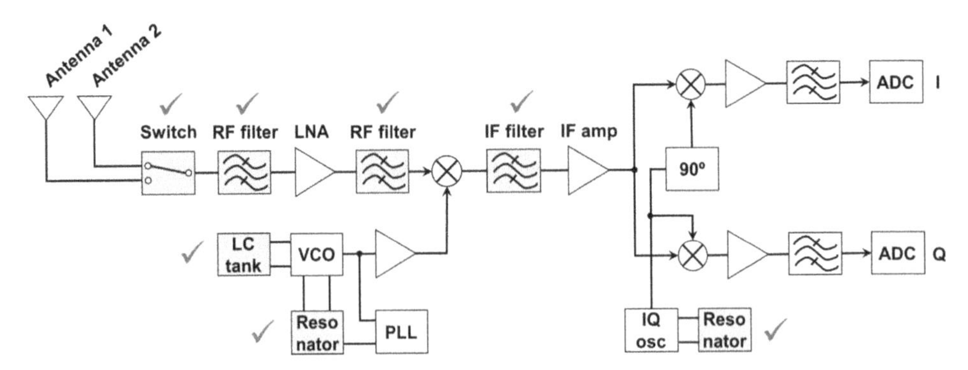

Figure 2.2. Super-heterodyne radio receiver block diagram [3]. Check-marked elements could be replaced with passive components based on RF-MEMS technology, thus improving performance and characteristics of the whole system.

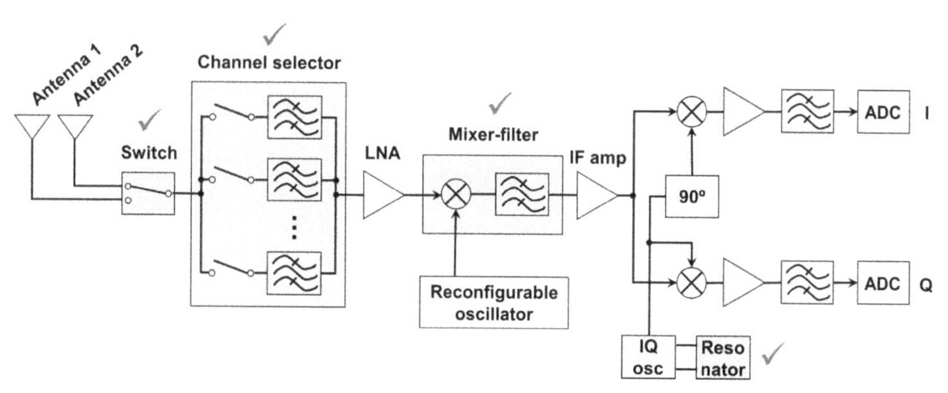

Figure 2.3. Super-heterodyne radio receiver block diagram, based on a modified architecture featuring complex blocks (check-marked) entirely implemented in RF-MEMS technology (i.e. multi-channel selector; reconfigurable oscillator; mixer-filter) as discussed in [3].

received signals that can be mixed and demodulated by the receiver, through the integration of a mixer-filter IF block also based on RF-MEMS technology.

As is straightforward to figure out, transceivers are just one of the numerous RF applications upon which RF-MEMS were expected to exert significant impact. For example, in modern radar systems the electronic antenna beam steering has replaced the old rotary mechanical antennas, leading to a significant reduction of complexity and space occupation, as well as to improved robustness of the whole system [10]. In spite of these advantages, phase shifters in standard technology (necessary to enable dynamic orientation of the antenna beam) are typically quite lossy. Therefore, redundant PAs, duplicated per each of the antenna array delay lines, must be inserted. A block diagram of the radar system, close to the transmitting antennas array and featuring standard phase shifters, is depicted in figure 2.4 (a) and discussed in [11].

Evidently, a PA is necessary before and after the phase shifter element on each branch, in order to regenerate the attenuated RF signal. For mid-power radar systems, traditional phase shifters could be replaced by high-performance low-loss realizations in RF-MEMS technology. In this case, power consumption and hardware complexity of the system could be considerably reduced, since a unique amplifying stage on each branch would be sufficient, as reported in figure 2.4 (b).

Besides the few examples discussed here in detail, RF-MEMS were expected to secure successful exploitation in many other contexts, with reference to the main fundamental market segments of telecommunications (both mobile handsets and infrastructure), industrial appliances, as well as aerospace and defence applications.

2.3 Fluctuating RF-MEMS market forecasts

In light of what has just been deployed in the previous section, RF-MEMS technology, especially in the first years of its experimental demonstration, was definitely not lacking ideas, perspectives and visions around its successful employment within commercial applications and, therefore, concerning market penetration. This way of interpreting the possibilities offered by MEMS for RF passives is

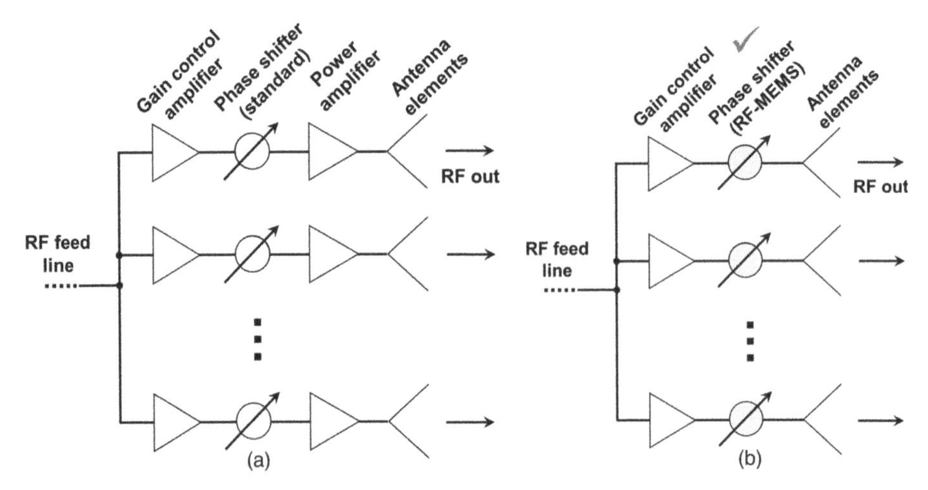

Figure 2.4. Block diagram scheme of a radar system employing electronically steerable antenna beams [11]. (a) Scheme employing phase shifters in standard technology. Power Amplifiers (PAs) are necessary on each branch in order to regenerate the signal attenuated by lossy phase shifting elements. (b) Scheme employing phase shifters in RF-MEMS technology. Due to the very low-loss of such components, the power consumption and hardware complexity of the system can be significantly reduced.

confirmed and corroborated by the first market forecast releases, signed by influential analysts who specialized in market observations. However, stimulating hints to deepen such an analysis emerge if one looks at the progression of the mentioned forecasts, keeping as a timeframe the first decade after RF-MEMS started to appear in the scientific literature, i.e. from the first years of the 2000s.

In early analyses, released in 2004–2005, the market volume was expected to ripple between 700 $M (millions of US Dollars—USD) and 1000 $M in 2009 [12] (see figure 2.5 (a)). Stepping forward in time, in 2006 WTC Consulting predicted a downsized market volume of RF-MEMS for mobile applications fluctuating around 10 $M in 2009 and 70 $M in 2011 [13] (see figure 2.5 (b)). Later, in August 2010, IHS Inc. consolidated the RF-MEMS market figure to a few $M in 2009, and predicted a volume of 225 $M in 2014 [14] (see figure 2.5 (c)). Nonetheless, the same IHS Inc. in 2012 shrunk the forecast for the 2014 market to less than 100 $M [15] (see figure 2.5(d)). In addition, Yole Développement in 2013 estimated a market volume for RF-MEMS of around 50 $M in 2014 and of less than 350 $M in 2018 [16] (see figure 2.5 (e)).

Abstracting from the numbers and summarizing the data listed in a qualitative fashion, RF-MEMS market forecasts were (somehow) systematically overestimated, and expectations were disappointed for about a decade. Moreover, recalling the concept of the hype curve previously reported in figure 2.1, and keeping in mind the plots in figure 2.5, it is as if RF-MEMS were characterized by a double peak of inflated expectations, the first one taking place around 2004–2005, and the second displaying around 2010, as discussed in [17]. This seems quite peculiar if compared to how hype curves typically develop, and is intimately related to the factors that are going to be discussed in the following section.

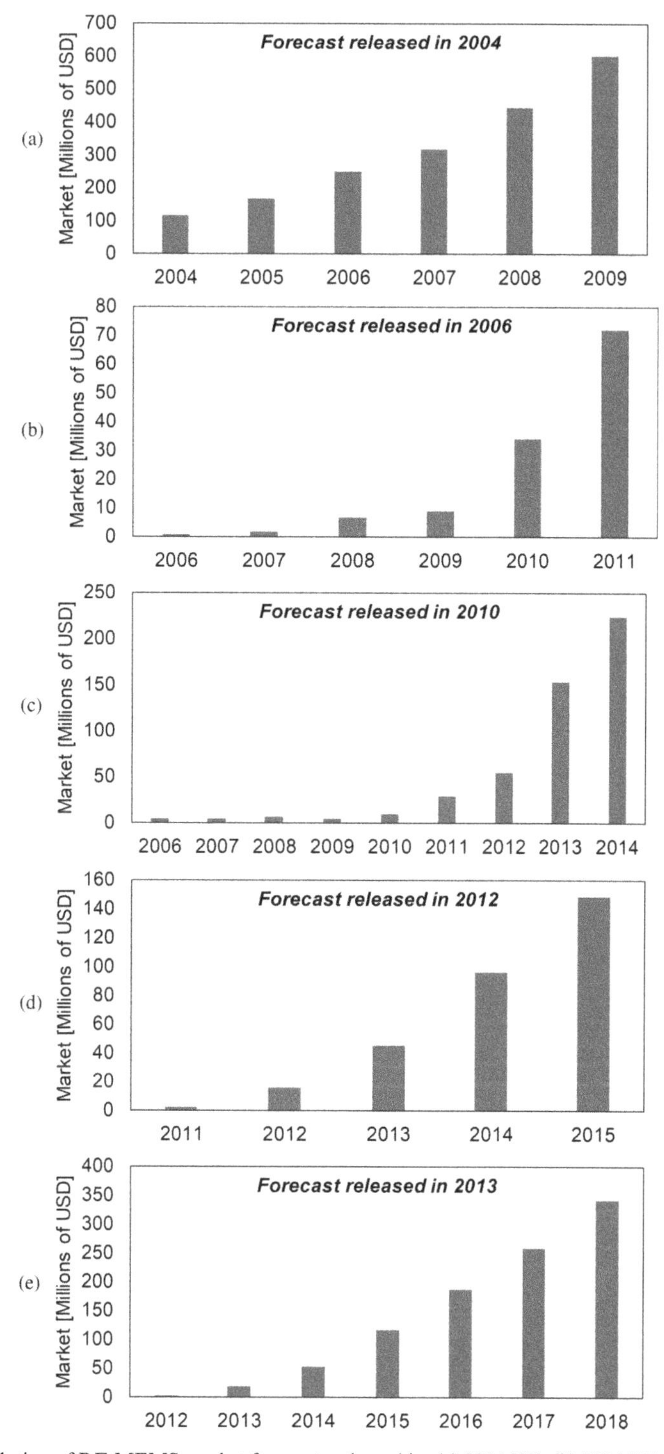

Figure 2.5. Evolution of RF-MEMS market forecasts released in: (a) 2004 [12]; (b) 2006 [13]; (c) 2010 [14]; (d) 2012 [15]; (e) 2013 [16].

2.4 Analysis of RF-MEMS market expectations (disappointment)

Providing a unique and straightforward interpretation to frame multiple fluctuations of RF-MEMS market forecasts is not a simple task. Nonetheless, in-depth analysis of the scientific literature and of its alternating trends over the years enables one to form sound explanations. That being said, the reasons for the repeatedly missed market achievements of RF-MEMS can be classified as belonging to two main categories, namely, *intrinsic* and *extrinsic*. The former factors are closely related to technology and functional aspects of RF passives based on microsystems. On the other hand, the latter are linked to the employment of RF-MEMS components within more complex sub-systems and systems. Referring again to the hype curve of RF-MEMS it can be inferred that intrinsic factors were the main cause for the drop of expectations after the first peak, taking place around 2006. Instead, extrinsic factors could be regarded as the primary reason for the second drop of expectations, right after 2010. Nonetheless, as explained in-depth in the following pages, extrinsic factors were present from the beginning, despite not being taken into proper consideration.

2.4.1 Intrinsic factors

Prior to enumerating the most relevant intrinsic factors causing a drop of expectations around market absorption of RF-MEMS, it is appropriate to aggregate the proper background in which they have to be framed. To this purpose, the concept of a development chain, tailored to the topic discussed in this work, is going to be discussed briefly. The latter is a schematic flow, not necessarily unique in its deployment, embracing all the phases and levels of intervention, bringing a certain design concept from inception to the physical demonstration and functional validation. A plausible implementation of the development chain for RF-MEMS is shown in figure 2.6.

A novel device idea has to be extensively modelled and simulated, in order to figure out the most suitable technology solutions to realize it and, after the process flow is chosen, one needs to optimize the RF-MEMS characteristics, bearing in mind

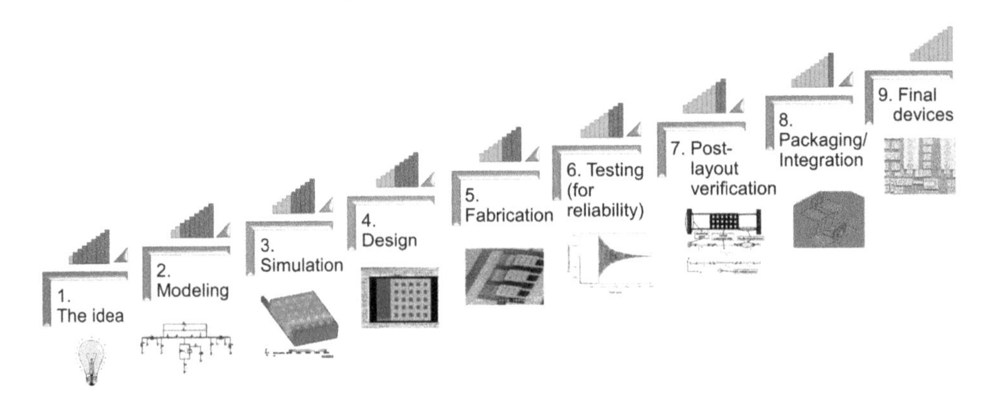

Figure 2.6. Typical arrangement of the RF-MEMS development chain that, through various steps, leads from inception of a novel design concept to the physical realization and experimental validation.

the most relevant existing trade-offs. Subsequently, all of the knowledge gathered through modelling and simulation is distilled into the layout of a physical device by means of the design phase, followed by the fabrication step. Once RF-MEMS specimens are available, they undergo lab-based experimental testing, in order to verify their operation and characteristics. Such a testing phase can be simply devoted to the functional verification of the fabricated samples, or it can be aimed at the reliability assessment of the Devices Under Test (DUTs). In the latter circumstance, the specifications imposed by the final application scenario of the RF-MEMS device are already known/defined, and DUTs are tested in such a way to verify if they are able to meet them or not. When lab-based experiments aim to verify RF-MEMS against final application requirements, this phase is typically addressed as testing for reliability. In general terms, when a novel design concept is manufactured and experimentally verified for the first time, the testing phase is more focused on the demonstration of the device functionality, rather than on the assessment of reliability, because of twofold motivation. On one hand, a few characteristics of the DUTs will differ to a certain extent from the nominal ones, targeted during the previous simulation step. Therefore, a redesign of improved layouts and additional fabrication batches will be necessary. On the other hand, when dealing with novel design concepts, it is plausible that neither the final application is clear in all its aspects, nor, in turn, the set of specifications to be targeted.

The testing phase is of particular relevance for the subsequent post-layout verification. During that, the set of predictive models, previously exploited for the initial modelling and simulation, are tuned based on the collected experimental datasets. This enables improved accuracy of the behavioural models that are then used to converge toward an optimized design, to be fabricated and tested again. It is straightforward that the more consolidated the exploitation of a certain technology is to realise RF-MEMS devices, the less extensive the effort has to be in the post-layout verification phase.

Progressing along the development chain, the subsequent phase concerns packaging and integration of RF-MEMS. Because of their complexity, such steps could be two individual phases of the whole chain. Moreover, both packaging and integration require additional validation and optimization cycles. Concerning the former, proper technology flow must be defined at first, and it has to be consolidated through subsequent fabrication runs, as happens for the RF-MEMS technology itself. In addition, issues related to the package-to-device interfacing should also be evaluated, both at the technology and functional levels. When dealing with integration instead, the most suitable solution, among those available, has to be identified, carefully evaluating all of the aspects that might exert a negative influence on the performance and reliability of the RF-MEMS devices.

As mentioned before, emerging in the previous discussion, the development chain is not characterized by a unique unrolling setting, even when referred to the same device. For instance, the level of technology maturity, as well as the expertise in designing certain classes of devices, exerts relevant impact on the effort to be invested in each single phase of the chain, as well as on the number of cycles and loops across a few stages, leading to optimized performance and characteristics.

However, single phases of the flow in figure 2.6, like the packaging, might require a standalone and ad-hoc development chain, if solid expertise is not already available.

Eventually, coming back to the initial discussion, the intrinsic factors that are going to be analysed in more detail are the following: *reliability*, *packaging* and *integration*. As a matter of fact, because they were not given proper consideration in the early days of MEMS for RF passives, they became one of the fundamental reasons triggering the first drop of expectations around market breakout of RF-MEMS, taking place in 2004 and 2005.

Reliability

Right after demonstration of the remarkable RF passives' characteristics achievable through exploitation of microsystem technology, reliability of RF-MEMS started to emerge as a major issue. In actuality, given their multi-physical behaviour, MEMS are exposed to a wide range of malfunctioning and failure mechanisms (both reversible and irreversible) that are quite common in material and mechanical engineering, but substantially unknown in the community of electronic and RF engineers [18]. As discussed in detail in [19], the most critical reliability issues at the functional level can be grouped as thermomechanical, electrical and environmental failures. They are briefly described as follows:

a) Contact wear (thermomechanical) of surfaces repeatedly in contact under conditions of large stresses (e.g. temperature and current densities) [20];
b) Fatigue (thermomechanical) in brittle and ductile materials resulting in progressive load bearing decrease, eventually bringing catastrophic failures [21];
c) Hardening (thermomechanical) of ductile materials (e.g. metals and alloys) taking place when overstress above the yield limit is reached [19];
d) Delamination (thermomechanical) of deposited layers occurring both because of processing defects or interfacial high stress levels [22];
e) Creep (thermomechanical) of metal layers undergoing time dependent loading at elevated temperatures [23];
f) Dielectric charging (electrical) due to the presence of large electric fields across thin dielectric films leading to accumulation of charges [24] that induce voltage screening [25] and/or stiction [26] (i.e. missed release of the switch to the rest position when the DC bias is removed);
g) Electromigration (electrical) leading to the formation of voids or hillocks due to high current densities in thin-film conductors [27];
h) Electrostatic Discharge (electrical) [28], dielectric breakdown [29], corona effects, and other electrically-driven irreversible failures due to very-small gaps typical of MEMS;
i) Micro-welding (electrical) due to accumulation of high current densities in very small areas, leading to formation of heat-induced welds of metals and joints [30];
j) Self-heating (electrical) caused by large currents and/or RF power flows through the MEMS device [31]. Self-heating favours the occurrence of thermomechanical failure mechanisms previously mentioned, like contact wear and creep;

k) Self-actuation (electrical) due to large RF power delivered to MEMS devices, able to generate an effective (spurious) DC voltage larger than pull-in threshold [32]. Failure modes labelled as **i)**, **j)** and **k)** fall under the denomination of power handling [33];
l) Corrosion (environmental), depending on the MEMS operation environmental characteristics (e.g. moisture, microorganisms, contaminants), and leading to a variety of failures [19];
m) Humidity (environmental), leading to various degradation effects and failures, among which is stiction induced by capillary forces acting on the surfaces in contact [34];
n) Radiation (environmental) in harsh environments like space, inducing modifications of device properties, spanning from reversible/irreversible drifts to catastrophic failures [35–36].

Undoubtedly, in the beginning the interest of the research community was mainly captured by demonstration of the remarkable characteristics of RF-MEMS versus standard semiconductor components. However, limited effort was dedicated to the reliability aspects listed above. Nevertheless, as soon as compliance of RF-MEMS devices started to be verified against sets of specifications originating from real applications, the demand for such a technology for further development aimed towards improving reliability emerged in a forceful fashion. Adopting this need as a fundamental driver, significant work was carried out at research and development level in subsequent years, which aimed to address one or more of the aforementioned failure mechanisms. As a result, lifetime, operability and medium-/long-term performance stability of RF-MEMS components were significantly improved. To this purpose, table 2.1 summarizes a few relevant solutions proposed and discussed in the literature to improve robustness against specific reliability issues.

As mentioned before, reliability was not the only aspect to be underestimated in the early stages of investigation of RF-MEMS technology. Packaging and integration are now going to be discussed.

Packaging

Also cross-linked related to reliability, the issues of packaging and encapsulation caused the initial enthusiasm triggered by RF-MEMS to be reshaped. As MEMS devices are composed by movable micro-membranes, they are very fragile and exposed to harmful environmental factors, such as mechanical shocks, moisture, dust particles and various contaminants. Therefore, RF passives in microsystem technology need to be properly isolated from the surrounding environment by being housed within a protective (possibly hermetic or semi-hermetic) capping [48–50]. In the case of RF-MEMS, the application of a package increases, on one hand, the complexity of technology and the manufacturing cost. Just to help build a more circumstanced idea around the impact of the package, in one of the first analyses it was estimated to be as high as 80% of the final product price [51]. On the other hand, the presence of a protective cap puts the outstanding RF performance of MEMS based passives in jeopardy, due to the longer paths that the signal has to travel across, increased parasitic

Table 2.1. Summary of a few relevant solutions proposed in the literature to improve the reliability of various RF-MEMS devices versus specific failure mechanisms and performance drifts.

Target device	Addressed failure mode/s	Solution	Ref.
Resonator	Long-term stability of resonant frequency	Low-temperature processing of Silicon thin-films	[37]
Lateral comb-drive resonator	Performance drifts driven by aging, device defects, harsh conditions, etc.	Custom active adaptive controller	[38]
MEMS with cyclic contact	Contact wear and stiction	Atomic Layer Deposition (ALD) of thin-film coatings	[39]
Resonator in a vacuum	Q-factor decrease due to degradation of in-package vacuum conditions	Use of getter films within the sealed cavity to absorb gases	[40]
Capacitive switch	Charge accumulation in the dielectric film, voltage screening and failure for stiction	Deposition of SiO_2/Si_3N_4 double layer dielectric stacks instead of single layer	[41]
Ohmic switch	Stiction failure due to micro-welding induced by large current densities	Active restoring mechanism (embedded micro-heater) activated in case of stiction	[42]
Capacitive switch	Stiction due to charge accumulation	Toggle design enabling push/pull active electrostatic control	[43]
Single Pole Single Throw (SPST)	Contact wear and other mechanical failures induced by cycled abrupt pull-in contact	Shaping of the biasing waveform to enable soft-landing/-contact of surfaces at pull-in	[44]
Capacitive switch	Contact wear and other mechanical failures induced by cycled abrupt pull-in contact	Resistive/capacitive braking schemes via design of embedded loading resistor and capacitor plate geometry	[45]
Lateral Bulk Acoustic Resonator (LBAR)	Resonant frequency drift due to temperature variations	Temperature compensation active control circuitry	[46]
Resonator	Resonant frequency drift due to temperature variations	Temperature Coefficient of Frequency (TCF) reduction by heavy n-type and p-type doping	[47]

effects, discontinuities, and so on. Therefore, the package should be carefully designed and accounted for as an actual part of the device from the beginning of development, thus making the design and modelling phases more challenging [52–55].

To complete the discussion on the packaging of RF-MEMS, the widely employed technique of Wafer-Level Packaging (WLP) is shown schematically in figure 2.7, and discussed below as a relevant example for explanatory purposes.

The WLP solution employs an entire wafer for the capping part, of the same dimension as the RF-MEMS device wafer to be packaged. The cap needs a proper scheme of signal redistribution paths that must correspond to the electrical pads of each MEMS device. The cap and device wafers must be carefully aligned to each other, as reported in figure 2.7 (a). A cross-sectional schematic view of what happens at the single device level is shown in figure 2.7 (b). Through Wafer Vias (TWVs) for signal redistribution are visible. In factual terms, TWVs are holes etched across the whole wafer thickness, then filled with metal, intended to contact RF-MEMS signal pads. Optionally, a shallow recess can be etched in the cap wafer backside, in order to yield housing for the RF-MEMS movable elevated parts. After bonding, the capping and device wafers come in to contact with each other, as depicted in figure 2.7 (c). Physical adhesion between the two parts is ensured by a reflow of solder bumps (see figure 2.7 (d)), which realize at the same time electrical interconnection between the RF-MEMS in-package devices and the external world.

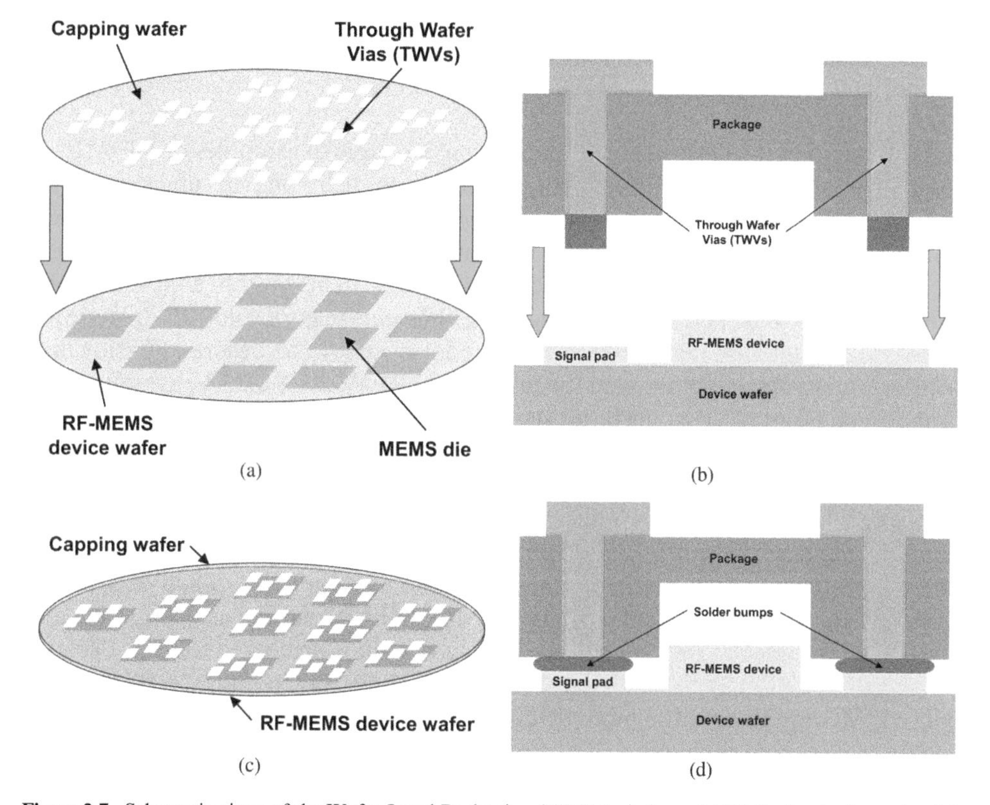

Figure 2.7. Schematic views of the Wafer-Level Packaging (WLP) technique. (a) Wafer-level schematic of the RF-MEMS device wafer and capping wafer alignment with Through Wafer Vias (TWVs) for signals redistribution. (b) Device-level cross-sectional schematic of the alignment between the RF-MEMS device and the package (TWVs are visible). (c) Wafer-level schematic of the aligned and bonded RF-MEMS device and capping wafers. (d) Device-level cross-sectional schematic of an in-package RF-MEMS device. Reflow of solder bumps is performed to ensure both electrical interconnections and physical adhesion between the device wafer and the package.

Once wafer-to-wafer adhesion is established, the packaged RF-MEMS wafer is ready to be cut into pieces (singulation), thus making available chips (containing one or a limited number of RF-MEMS devices) ready to be mounted, i.e. integrated, in a more complex system or sub-system.

Unquestionably, WLP, albeit commonly employed, is not the only solution to package RF-MEMS. Other techniques and technologies exist, with a wide deck of advantages and disadvantages with respect to WLP. Such solutions are not reported here, as their discussion falls out of the scope of this work. Nonetheless, they are extensively covered by other worthy contributions easily traceable in the scientific literature [56–58].

Integration

Finally, yet importantly, MEMS technology is typically incompatible with standard semiconductor platforms, such as CMOS. Consequently, in-package RF-MEMS passive components need to be integrated with active electronics at sub-systems or system level, i.e. on a board. This step is typically performed through Surface Mount Technologies (SMTs), and ad-hoc circuitry must be developed and deployed as well, in order to operate/drive RF-MEMS, raising, also in this circumstance, complexity and costs [59–64].

As for the previous discussion on packaging, a couple of examples of RF-MEMS integration are briefly reported next, in order to help one to understand the multitude of issues and hurdles related to such a delicate phase of the development chain. The case studies mentioned are shown schematically in figure 2.8. In both circumstances, integration of the in-package RF-MEMS device into a more complex system is meant to be performed through SMT. To this regard, solder balls deployed on the package front side are meant to undergo reflow to ensure adhesion and electrical interconnection between the in-package RF-MEMS and the system. From a conceptual point of view, the latter step is equivalent to the reflow of solder bumps previously reported in figure 2.7 (d) to establish adhesion and electrical continuity between the RF-MEMS device wafer and the package.

Nonetheless, apart from ensuring in-package RF-MEMS integration, the two examples reported in figure 2.8 realize an additional functionality, albeit through different strategies. As a matter of fact, in both cases the CMOS active circuitry, necessary for controlling and driving the RF-MEMS device, is integrated and interfaced through the package, thus carrying out the so-called System in Package (SiP) strategy, despite in this case the system order of complexity being quite reduced. Moving into a little more detail, in the example reported in figure 2.8 (a) a matrix of electrical interconnections is designed and realized on the RF-MEMS device wafer. The CMOS chip is mounted upside-down upon such contacts thanks to a through-wafer recess etched in the package, meant to provide proper housing. The adhesion and electrical continuity between the CMOS chip and the underlying pads is ensured by a reflow of solder balls. However, in the example reported in figure 2.8 (b), the recess etched in the cap is shallow and does not develop across the whole wafer thickness. In this case, the CMOS chip is placed into the housing with

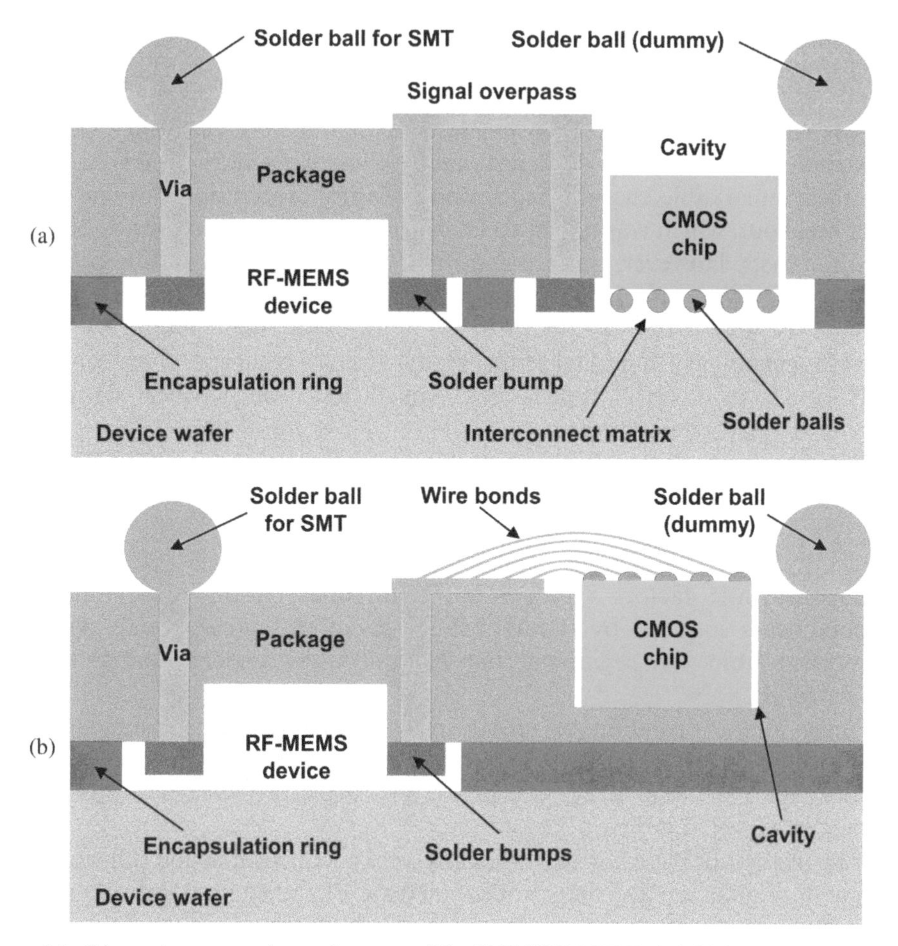

Figure 2.8. Schematic cross-sections of two possible CMOS/RF-MEMS hybrid in-package integration solutions, for Surface Mounting Technology (SMT). (a) The electrical interconnection matrix is deployed on the RF-MEMS device wafer. A through-wafer recess in the package allows housing of the CMOS chip. (b) The CMOS chip is housed within a shallow recess in the package. Electrical interconnections are ensured by wire bonding.

the contacts upward. The electrical interconnection between the CMOS driving circuitry and the RF-MEMS is ensured by wire bonds.

In conclusion, bearing in mind the previous sections, it is straightforward that the readiness of RF-MEMS technology for market absorption was still far from maturity in the first years of its investigation. In any event, once the scientific community became aware of the aforementioned issues, plenty of effort was spent on them, thus significantly reducing the gap of RF-MEMS in terms of reliability, encapsulation and integration/compatibility with standard semiconductor active technologies, and boosting, in turn, expectations and optimism at the market level for years to come.

Unfortunately, RF-MEMS technology still had impediments to overcome, as will be discussed in the following section.

2.4.2 Extrinsic factors

The frame of reference depicted in the previous pages appears to be significantly complex. The just-reviewed critical intrinsic factors, namely, reliability, packaging and integration, were clearly underestimated in the early days of RF-MEMS, in which their remarkable characteristics were inflating expectations around massive market breakout, which was meant to be imminent, at least according to the initial market forecasts. However, the massive effort brought by the scientific community in the following years drove paramount improvement of RF-MEMS technology against all the weaknesses mentioned in the previous sections; making RF passives in MEMS technology, in factual terms, ready—from a technical point of view—to be employed in commercial applications. Nonetheless, what has been discussed up to here is just part of the picture.

If the discussion of intrinsic factors impairing the success of RF-MEMS was complex and rich in ramifications, the one around extrinsic factors is much simpler. However, this does not mean that this class of factors exerted less influence on the missed penetration of MEMS-based RF passives into the market. In fact, most probably, extrinsic factors represent the substantial reason for the thread of disappointments collected by RF-MEMS across more than a decade, since they were present from the beginning, but fundamentally underestimated, as were intrinsic factors.

Looking at the evolution of mobile phones' communication standards [65], starting from the early 1980s with 1G (i.e. first generation) to the latest 3.5G and 4G across the first and second decade of the 2000s, the scenario appears quite extended, as reported in figure 2.9.

Across the decades, crucial steps forward were taken, such as the transition from analogue to digital signals, the seamless increase of transmitted data volumes, as well as the increase of other-than-voice services, venturing into the field of multi-media exchange. In parallel to the evolution of communication standards and to the increase of services, mobile handsets followed a seamless path in size reduction (as shown in figure 2.10), weight, power consumption and costs.

The crystal-clear fact of this scenario is that the evolution just discussed (for certain aspects similar to a revolution) and reported in figure 2.9 and figure 2.10, took place without RF-MEMS. In other words, in the huge consumer market segment of mobile phones, there has never been a strong enough need for RF passives with boosted performance, prompting the investigation of technologies other than standard ones, opening up the floor to issues such as reliability, packaging and integration, as well as to higher costs for the final single component.

In conclusion, the combination of intrinsic and extrinsic factors that caused multiple drops of inflated expectations in the hype curve, led the scientific community (circa 2010) to the unspoken opinion that RF-MEMS technology was mainly meant for niche and small volume applications, such as in the space and defence sector. In the latter, indeed, the driver of remarkable performance rules over increased price and technical hurdles at the integration level.

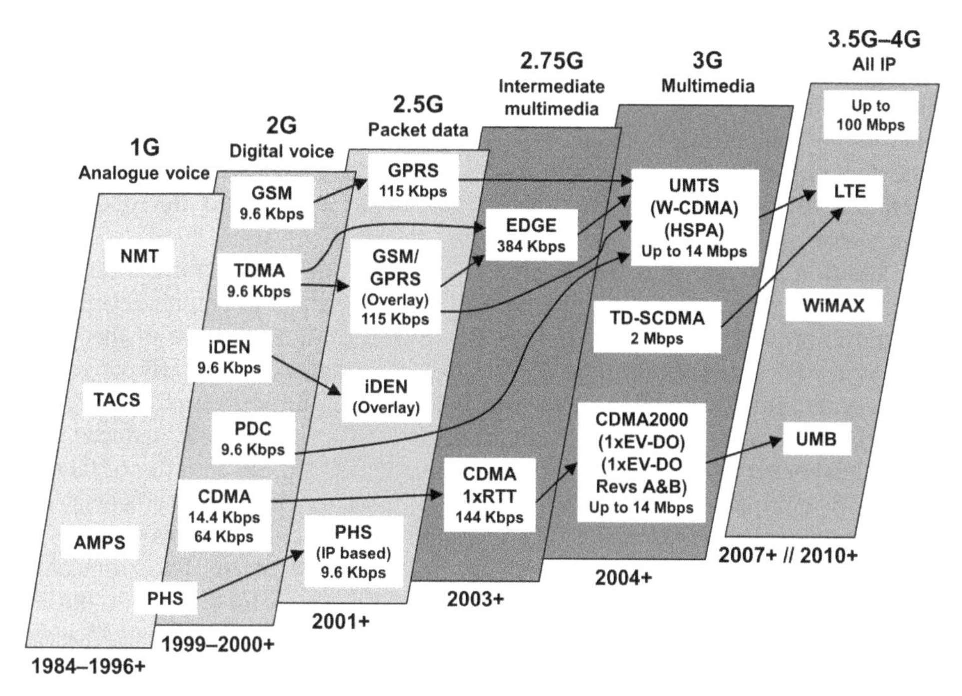

Figure 2.9. Evolution of mobile communication standards from 1G, in the early 1980s, to 3.5G and 4G, across the first and second decade of the 2000s [65].

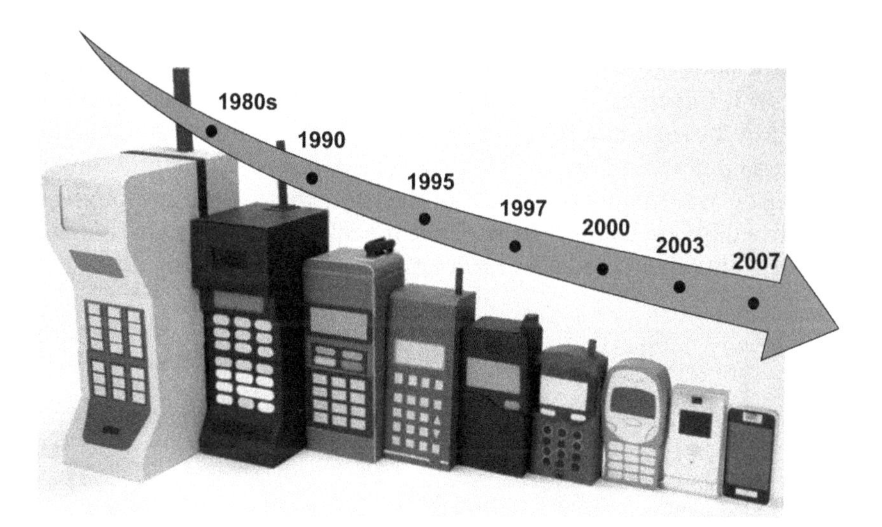

Figure 2.10. Schematic view of the seamless trend in size reduction of mobile handsets, from the 1980s to the first decade of the 2000s.

2.5 The RF-MEMS market today

Quite unexpectedly, the market segment of smartphones, which started to expand massively in a few years, brought in some elements partially altering the landscape described in the previous pages. With the widespread diffusion of mobile phones in the mid-1990s, the quality of the voice signal has been experiencing a degradating trend, hopping from one generation to the next (i.e. GSM, 2.5G, and so on), as stated in [66] and shown in figure 2.11. The increasing use of cellular communications, the introduction of full-screen devices (with touch technology) and the integration of the antenna inside the handset, have determined this decrease in performance.

In addition, as smartphones became smaller and thinner, it turned out to be difficult to integrate proper circuitry to boost performance, to counteract dropped calls and improve voice quality. In [66] it is estimated that the ratio of theoretical versus actual RF signal quality was decreasing with a pace of about 1 dB per year for over a decade. In light of all these factors, the antennas of recent smartphones were not working efficiently anymore, leading to slower download speeds, reduced voice quality, lower energy efficiency and more dropped calls [67]. As a matter of fact, the demand for compact and device integrated antennas decreased the efficiency and made them more sensitive to interaction with external items, such as the head of the speaker or the hand holding the device. In light of this scenario, fixed impedance matching between the antenna and the RF Front End (RFFE) was not an optimal solution anymore, as the characteristic impedance of the transmitting/receiving aerial undergoes slight changes.

The availability of high-performance reconfigurable impedance matching tuners started to be desirable, and RF-MEMS were shown to be a proper technology solution to implement such complex devices. In 2012, information about the employment of RF-MEMS adaptable impedance tuners manufactured by WiSpry within the Samsung Focus Flash Windows smartphone was announced [68]. Later, in autumn 2014, Cavendish Kinetics (CK) announced the commercial adoption of its RF-MEMS based SmarTune^{TM} antenna tuning solution in the Nubia Z7 smartphone, manufactured by the Chinese ZTE Corporation [69]. A few months later, CK again went public with the information that SmarTune products were adopted in five different Chinese smartphone models, promising signal enhancement of 2–3 dB, resulting in boosted data rates (up to 2×) and improved battery life (up to

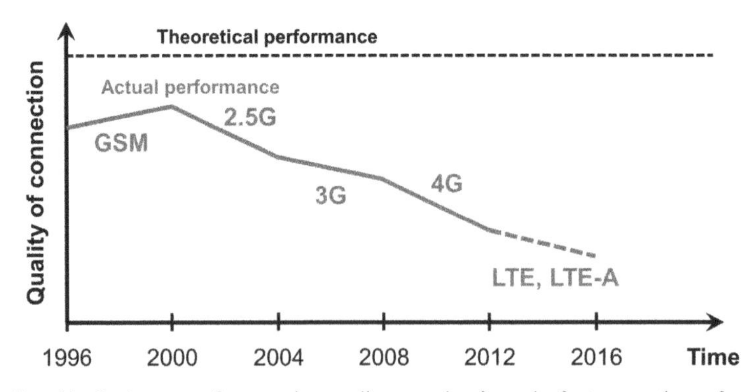

Figure 2.11. Trend in the decrease of connection quality, stepping from the first generations of mobile handsets in the mid-1990s, to modern 4G-LTE smartphones [66].

40%) [70]. To demonstrate that this was not just a flash in the pan, Qorvo has more than quadrupled its RF-MEMS sales in the last three years from 145 $M to 585 $M, as reported in [71]. Such a remarkable achievement has been reached by extensively integrating RF-MEMS switches within RFFEs, in order to extend their reconfigurability and, in turn, their operability. This brought the same hardware to communicate simultaneously and in an effective fashion, over Wi-Fi, Bluetooth, and cellular bands belonging to 3G and 4G networks.

Bearing in mind the overall discussion developed throughout this chapter, the most sceptical observers might call into question that the just-listed examples are simply another peak of inflated expectations preceding the next (and probably last) drop of enthusiasm in the hype curve of RF-MEMS. Nonetheless, a fundamental element of discontinuity with respect to the past must be taken into careful consideration. Starting from the exacerbation of the negative trend discussed in figure 2.11, the surge in market need for RF passives with improved characteristics and boosted reconfigurability was unmistakable. In other words, recalling what was previously discussed in this chapter, the approach to exploitating RF-MEMS began to rotate from a mainly technology push scenario, to a market driven philosophy.

As a conclusive remark, coming back for a moment to the vision of Nguyen [3] (previously reviewed in figure 2.2 and figure 2.3), it looks like it is somehow taking place, despite a reversal in development. The first market exploitation of RF-MEMS technology is, in fact, related to complex reconfigurable networks (impedance tuners at first), which was supposed to be the second phase of RF passives in MEMS technology (see figure 2.3). The spreading of these networks has been increasing the demand for high-performance, reliable and low-cost basic RF-MEMS components (mainly switches and tunable capacitors), thus pushing advanced research and engineering of such building blocks as well.

Conclusion

This chapter studied, in a quite detailed fashion, the evolution of the RF-MEMS market, from the early days in which the technology started to be discussed within the scientific community. As a first step, critical concepts of market analysis were introduced, in order to support the subsequent discussion. In detail, these tools of abstraction are the hype curve and technology push versus market pull/driven scenarios. Briefly, the hype curve sketches the behavioural up and down trend versus time of market expectations fed by a novel technology, since the first discussion in the scientific community, to full maturity. However, the technology push and market pull are simply two different formulations of the so-called *value proposition* related to a product realized through a certain technique and/or technology. In the technology push approach, a strong market need for addressing a certain problem or issue, in fact, does not exist. Therefore, the value has leverage upon different drivers, such as, for instance, the added value in terms of distinctiveness related to a certain solution. In contrast, in a market pull/driven scenario, one or more market needs clearly exist, and the added value of a given technology solution resides right within its capability of addressing those very requirements.

After reviewing such fundamental concepts, they were applied to the case study focus of this book. The vision of the early days of RF-MEMS, concerning their expected penetration into the mobile handsets market, was reported. It was envisioned to act in a first phase at single components level. In other words, in standard transmitter/receiver (transceiver) architectures, RF-MEMS simple devices, like switches and variable capacitors (varactors), were meant to replace standard devices. Later on, more complex RF passive networks in MEMS technology, like switching matrices and tunable filters, would have been the pivot around which transceivers architecture was redesigned, aiming for better performance and improved reconfigurability.

As a matter of fact, all of the most positive market forecasts, released from the early years of the 2000s for more than a decade, resulted in hope being systematically disappointed. The most relevant causes leading to such a tepid success of RF-MEMS technology were identified, splitting into two main classes, namely intrinsic and extrinsic factors. The former were intimately linked to aspects of the technology that were not properly accounted for in the beginning, and were identified to be reliability, packaging and integration of RF-MEMS. Such items have been analysed with an in-depth technical insight. The latter, instead were related to scarce readiness of market needs with respect to the boosted characteristics offered by RF-MEMS solutions, albeit counterbalanced by additional issues at integration level and higher costs.

Finally, the environmental changes taking place in the market in recent years were also mentioned and discussed. Such modifications brought the approach to exploit RF-MEMS from a less favourable technology push scenario, to a more advantageous market pull reference frame. The most recent success stories related to RF-MEMS products employed in consumer applications were listed, proving the achievement and consolidation of an appropriate level of maturity. This will constitute the foundation for the future spread of RF-MEMS devices, which will be discussed in the following pages of this work.

References

[1] Gartner Inc. *Interpreting Technology Hype*http://www.gartner.com/technology/research/methodologies/hype-cycle.jsp accessed 29 May 2017

[2] Martin M J C 1994 *Managing Innovation and Entrepreneurship in Technology-Based Firms* 1st edn (New York: John Wiley) p 416

[3] Nguyen C T-C 2001 Transceiver front-end architectures using vibrating micromechanical signal processors. *Proc. Top. Meet. on Silicon Monolithic Integrated Circ. in RF Sys. (Ann Arbor, MI, Sept. 2001)* pp 23–32

[4] Nguyen C T-C 2002 RF MEMS for wireless applications *Proc. Device Res. Conf. DRC (Santa Barbara, CA, June 2002)* pp 9–12

[5] Laskar J, Matinpour B and Sudipto C 2004 *Modern Receiver Front-Ends: Systems, Circuits, and Integration* 1st edn (New York: John Wiley) p 221

[6] Nguyen C T-C 2006 Integrated micromechanical circuits for RF front ends *Proc. Europ. Solid-State Circ. Conf. ESSCIRC (Montreux, Sept 2006)* pp 7–16

[7] Nguyen C T-C 1998 Microelectromechanical devices for wireless communications. *Proc. Int. Workshop on Micro Electro Mechan. Syst. MEMS (Heidelberg, Jan. 1998)* pp 1–7
[8] Nguyen C T-C 2007 MEMS technology for timing and frequency control *IEEE Trans. Ultrason. Ferroelectr. Freq. Control* **54** 251–70
[9] Nguyen C T-C 2013 MEMS-based RF channel selection for true software-defined cognitive radio and low-power sensor communications *IEEE Commun. Mag.* **51** 110–9
[10] Meikle H 2001 *Modern Radar Systems* 1st edn (Norwood: Artech House) p 563
[11] Haridas N, Erdogan A T, Arslan T, Walton A J, Smith S, Stevenson T, Dunare C, Gundlach A, Terry J, Argyrakis P, Tierney K, Ross A and O'Hara T 2008 Reconfigurable MEMS antennas *Proc. NASA/ESA Conf. on Adaptive Hardware and Syst. (Noordwijk, June 2008)* pp 147–54
[12] Bouchaud J, Knoblich B and Wicht H 2006 RF MEMS market *Proc. German Microw. Conf. GeMiC (Karlsruhe, March 2006)* pp 1–2
[13] Johnson R C*RF MEMS Aim for Cell Phone on-a-Chip* http://www.memsjournal.com/2010/07/rf-mems-aims-for-cell-phone-onachip.html accessed 31 May 2017
[14] MEMS Journal *Cell Phone Antenna Troubles? RF MEMS Come to the Rescue* http://www.memsjournal.com/2010/09/cell-phone-antenna-troubles-rf-mems-come-to-the-rescue.html accessed 31 May 2017
[15] McGrath D *Teardown Finds RF MEMS in Samsung Handset* http://www.eetimes.com/document.asp?doc_id=1260917 accessed 31 May 2017
[16] DeLisle J-J *The Future Of Connectivity: Mobile & Automobiles* http://www.mwrf.com/systems/future-connectivity-mobile-automobiles accessed 31 May 2017
[17] Iannacci J 2015 RF-MEMS: an enabling technology for modern wireless systems bearing a market potential still not fully displayed *Microsyst. Tech.* **21** 2039–52
[18] Iannacci J 2014 Reliability of MEMS: A perspective on failure mechanisms, improvement solutions and best practices at development level *Displays* **37** 62–71
[19] Hartzell A L, da Silva M G and Shea H 2011 *MEMS Reliability* 1st edn (New York: Springer) p 291
[20] Chianrabutra C, Jiang L, Lewis A P and McBride J W 2013 Evaluating the influence of current on the wear processes of Au/Cr-Au/MWCNT switching surfaces *Proc. Holm Conf. on Electrical Contacts (Newport, RI, Sept. 2013)* pp 1–6
[21] Kawai T, Gaspar J, Paul O and Kamiya S 2009 Prediction of strength and fatigue lifetime of MEMS structures with arbitrary shapes *Proc. Solid-State Sens., Actuators and Microsystems Conf. TRANSDUCERS (Denver, CO, June 2009)* pp 1067–70
[22] Khanna V K 2011 Adhesion–delamination phenomena at the surfaces and interfaces in microelectronics and MEMS structures and packaged devices *J. Phys. D: Appl. Phys.* **44** 1–19
[23] Van Gils M, Bielen J and McDonald G 2007 Evaluation of creep in RF MEMS devices *Proc. Int. Conf. on Thermal, Mech. and Multi-Physics Simulation Experiments in Microelectron. and Micro-Syst. EuroSime (London, April 2007)* pp 1–6
[24] Tavassolian N, Koutsoureli M, Papaioannou G, Lacroix B and Papapolymerou J 2010 Dielectric charging in capacitive RF MEMS switches: The effect of electric stress *Proc. Asia-Pacific Microw. Conf. APMC (Yokohama, Dec. 2010)* pp 1833–6
[25] Iannacci J 2013 *Practical Guide to RF-MEMS* 1st edn (Weinheim: Wiley) p 372
[26] Zaghloul U, Bhushan B, Pons P, Papaioannou G, Coccetti F and Plana R 2011 Different stiction mechanisms in electrostatic MEMS devices: Nanoscale characterization based on adhesion and friction measurements *Proc. Int. Solid-State Sensors, Actuators and Microsystems Conf. TRANSDUCERS (Beijing, June 2011)* pp 2478–81

[27] Wilson C J, Horsfall A B, O'Neill A G, Wright N G, Bull S J, Terry J G, Stevenson J T M and Walton A J 2007 Direct Measurement of electromigration-induced stress in interconnect structures *IEEE Trans. Dev. Mater. Reliab.* **7** 356–62

[28] Sangameswaran S, De Coster J, Linten D, Scholz M, Thijs S, Haspeslagh L, Witvrouw A, Van Hoof C, Groeseneken G and De Wolf I 2008 ESD reliability issues in microelectromechanical systems (MEMS): A case study on micromirrors *Proc. Electrical Overstress/Electrostatic Discharge Symp. EOS/ESD (Tucson, AZ, Sept. 2008)* pp 249–57

[29] Tazzoli A, Barbato M, Ritrovato V and Meneghesso G 2010 A comprehensive study of MEMS behavior under EOS/ESD events: Breakdown characterization, dielectric charging, and realistic cures *Proc. Electrical Overstress/Electrostatic Discharge Symp. EOS/ESD (Reno, NV, Oct 2010)* pp 1–10

[30] Tazzoli A, Iannacci J and Meneghesso G 2011 A positive exploitation of ESD events: Micro-welding induction on ohmic MEMS contacts *Proc. Electrical Overstress/Electrostatic Discharge Symp. EOS/ESD (Anaheim, CA, Sept. 2011)* pp 1–8

[31] Ivira B, Fillit R-Y, Ndagijimana F, Benech P, Parat G and Ancey P 2008 Self-heating study of bulk acoustic wave resonators under high RF power *IEEE Trans. Ultrason. Ferroelectr. Freq. Control* **55** 139–47

[32] Malmqvist R, Jonsson R, Samuelsson C, Reyaz S, Rantakari P, Ouacha A, Vaha-Heikkila T, Varis J and Rydberg A 2012 Self-actuation tests of ohmic contact and capacitive RFMEMS switches for wideband RF power limiter circuits *Proc. Int. Semiconductor Conf. CAS (Sinaia, Oct. 2012)* pp 217–20

[33] Blondy P and Peroulis D 2013 Handling RF power: The latest advances in RF-MEMS tunable filters *IEEE Microw. Mag.* **14** 24–38

[34] Wu L, Noels L, Rochus V, Pustan M and Golinval J-C 2011 A Micro–macroapproach to predict stiction due to surface contact in microelectromechanical systems *IEEE J. Microelectromechan. Syst.* **20** 976–90

[35] Lozano A, Palumbo F and Alurralde F 2009 Radiation effects on SOI electrostatic comb drive actuators of MEMS devices *Proc. Argentine School of Micro-Nanoelectronics, Technology and Applications EAMTA (San Carlos de Bariloche, Oct. 2009)* pp 46–9

[36] Goldsmith C L, Hwang J C M, Gudeman C, Auciello O, Ebel J L and Newman H S 2012 Robustness of RF MEMS capacitive switches in harsh environments *Proc. IEEE MTT-S Int. Microwave Symp. (Montreal, June 2012)* pp 1–3

[37] Sousa P M, Chu V and Conde J P 2012 Reliability and stability of thin-film amorphous silicon MEMS resonators *J. Micromech. Microeng.* **22** 1–8

[38] Izadian A and Famouri P 2008 Reliability enhancement of MEMS lateral comb resonators under fault conditions *IEEE Trans. Control Syst. Technol.* **16** 726–34

[39] Ashurst W R, Jang Y J, Magagnin L, Carraro C, Sung M M and Maboudian R 2004 Nanometer-thin titania films with SAM-level stiction and superior wear resistance for reliable MEMS performance *Proc. IEEE Int. Conf. on Micro Electro Mechanical Systems MEMS (Maastricht, Jan. 2004)* pp 153–6

[40] Longoni G, Conte A, Moraja M and Fourrier A 2006 Stable and reliable Q-factor in resonant MEMS with getter film *Proc. IEEE Int. Reliab. Phys. Symp. IRPS (San Jose, CA, March 2006)* pp 416–20

[41] Li G, Hanke U, Cheng Z, Min D, San H and Chen X 2011 Si_3N_4/SiO_2 dielectric stacks for high reliable capacitive RF MEMS switch *Proc. IEEE-NANO Conf. on Nanotech. (Portland, OR, Aug. 2011)* pp 496–9

[42] Iannacci J, Faes A, Repchankova A, Tazzoli A and Meneghesso G 2011 An active heat-based restoring mechanism for improving the reliability of RF-MEMS switches *Microelectron. Reliab.* **51** 1869–73

[43] Solazzi F, Tazzoli A, Farinelli P, Faes A, Mulloni V, Meneghesso G and Margesin B 2010 Active recovering mechanism for high performance RF MEMS redundancy switches *Proc. Europ. Microw. Conf. EuMC (Paris, Sept. 2010)* pp 93–6

[44] Czaplewski D A, Dyck C W, Sumali H, Massad J E, Kuppers J D, Reines I, Cowan W D and Tigges C P 2006 A soft-landing waveform for actuation of a single-pole single-throw ohmic RF MEMS switch *IEEE J. Microelectromechan. Syst. JMEMS* **15** 1586–94

[45] Ankit J, Nair P R and Alam M A 2011 Strategies for dynamic soft-landing in capacitive microelectromechanical switches *Appl. Phys. Lett.* **98** 1–3

[46] Lavasani H M, Pan W, Harrington B P, Abdolvand R and Ayazi F 2012 Electronic temperature compensation of lateral bulk acoustic resonator reference oscillators using enhanced series tuning technique *IEEE J. Solid-State Circuits* **47** 1381–93

[47] Pensala T, Jaakkola A, Prunnila M and Dekker J 2011 Temperature compensation of silicon MEMS resonators by heavy doping *Proc. IEEE Int. Ultrasonics Symp. IUS (Orlando, FL, Oct. 2011)* pp 1952–5

[48] Jourdain A, Ziad H, De Moor P and Tilmans H A C 2003 Wafer-scale 0-level packaging of (RF-)MEMS devices using BCB *Proc. Symp. on Design, Test, Integ. and Packaging of MEMS/MOEMS DTIP (Mandelieu-La Napoule, May 2003)* pp 239–44

[49] Park Y-K, Park H-W, Lee D-J, Park J-H, Song I-S, Kim C-W, Song C-M, Lee Y-H, Kim C-J and Ju B K 2002 A novel low-loss wafer-level packaging of the RF-MEMS devices *Proc. IEEE Int. Conf. on Micro Electro Mechanical Systems (Las Vegas, NV, Jan. 2002)* pp 681–4

[50] Park Y-K, Kim Y-K, Hoon K, Lee D-J, Kim C-J, Ju B-K and Park J-O 2003 A novel thin chip scale packaging of the RF-MEMS devices using ultra thin silicon *Proc. IEEE Int. Conf. on Micro Electro Mechanical Systems (Kyoto, Jan. 2003)* pp 618–21

[51] Cohn M B, Roehnelt R, Xu J-H, Shteinberg A and Cheung S 2002 MEMS packaging on a budget (fiscal and thermal) *Proc. Int. Conf. on Electronics, Circ. and Systems ICECS (Dubrovnik, Sept. 2002)* pp 287–90

[52] Iannacci J, Tian J, Sosin S, Gaddi R and Bartek M 2006 Hybrid wafer-level packaging for RF MEMS applications *Proc. Int. Wafer-Level Packaging Conf. IWLPC (San Jose, CA, Nov. 2006)* pp 106–13

[53] Iannacci J, Bartek M, Tian J, Gaddi R and Gnudi A 2008 Electromagnetic optimization of an RF-MEMS wafer-level package *Sensors Actuators A* **142** 434–41

[54] Margomenos A and Katehi L P B 2002 DC to 40 GHz on-wafer package for RF MEMS switches *Proc. IEEE Top. Meet. on Electrical Performance of Electronic Packaging (Monterey, CA, Oct. 2002)* pp 91–4

[55] Margomenos A and Katehi L P B 2003 High frequency parasitic effects for on-wafer packaging of RF MEMS switches *Proc. IEEE MTT-S Int. Microw. Symp. (Philadelphia, PA, June 2003)* pp 1931–4

[56] Lau J H, Lee C, Premachandran C S and Aibin Y 2009 *Advanced MEMS Packaging* 1st edn (New York: McGraw-Hill) p 576

[57] Kuang K, Kim F and Cahill S S (ed) 2010 *RF and Microwave Microelectronics Packaging* 1st edn (New York: Springer) p 285

[58] Kuang K and Sturdivant R (ed) 2017 *RF and Microwave Microelectronics Packaging II* 1st edn (New York: Springer) p 172

[59] De Silva A P and Hughes H G 2003 The package integration of RF-MEMS switch and control IC for wireless applications *IEEE Trans. Adv. Packag.* **26** 255–60

[60] Lu A C W, Chua K M and Li H G 2005 Emerging manufacturing technologies for RFIC, antenna and RF-MEMS integration *Proc. IEEE Int. Workshop on Radio-Freq. Integr. Technol.: Integ. Circ. for Wideband Commun. and Wireless Sensor Networks (Singapore, Nov–Dec 2005)* pp 142–6

[61] Pacheco S, Zurcher P, Young S, Weston D and Dauksher W 2004 RF MEMS resonator for CMOS back-end-of-line integration *Proc. Top. Meet. on Silicon Monolithic Integ. Circ. in RF Systems (Atlanta, GA, Sept. 2004)* pp 203–6

[62] Th Rijks G S M, van Beek J T M, Ulenaers M J E, De Coster J, Puers R, den Dekker A and van Teeffelen L 2003 Passive integration and RF MEMS: a toolkit for adaptive LC circuits *Proc. Europ. Solid-State Circ. Conf. ESSCIRC (Estoril, Sept. 2003)* pp 269–72

[63] Zhang Q X, Yu A B, Yang R, Li H Y, Guo L H, Liao E B, Tang M, Kumar R, Liu A Q, Lo G Q, Balasubramanian N and Kwong D L 2006 Novel monolithic integration of RF-MEMS switch with CMOS-IC on organic substrate for compact RF system *Proc. IEEE Int. Electron. Dev. Meet. IEDM (San Francisco, CA, Dec. 2006)* pp 1–4

[64] Ziegler V, Siegel C, Schonlinner B, Prechtel U and Schumacher H 2005 RF-MEMS switches based on a low-complexity technology and related aspects of MMIC integration *Proc. Europ. Gallium Arsenide and other Semiconductor Application Symp. EGAAS (Paris, Oct. 2005)* pp 289–92

[65] Turner B and Orange M *3G tutorial* https://www.slideshare.net/9876012345/3g-tutorial-38911321 accessed 9 June 2017

[66] Allan R *RF MEMS Switches are Primed for Mass-Market Applications* http://www.mwrf.com/active-components/rf-mems-switches-are-primed-mass-market-applications accessed 9 June 2017

[67] Jacobs School of Engineering *RF MEMS: New Possibilities for Smartphones* http://jacobsschool.ucsd.edu/pulse/winter2014/page5.sfe accessed 9 June 2017

[68] IHS iSuppli *IHS iSuppli Teardown Analysis Service Identifies First Use of RF MEMS Part, Set to be Next Big Thing in Cellphone Radios* http://news.ihsmarkit.com/press-release/design-supply-chain/ihs-isuppli-teardown-analysis-service-identifies-first-use-rf-mems accessed 9 June 2017

[69] Cavendish Kinetics *Nubia Adopts Cavendish Kinetics SmarTuneTM Antenna Tuning Solution for its new Z7 LTE Smartphone* http://www.cavendish-kinetics.com/news/news-releases/ accessed 9 June 2017

[70] Cavendish Kinetics *Cavendish's SmarTune™ Antenna Tuning Solution now shipping in 5 smartphone models* http://www.cavendish-kinetics.com/release/cavendish-kinetics-adds-design-wins-and-ramps-shipments-of-rf-mems-tuners/ accessed 9 June 2017

[71] Morra J *Chip Makers Build Fortunes From RF MEMS* http://www.mwrf.com/semiconductors/chip-makers-build-fortunes-rf-mems?NL=MWRF-001&Issue=MWRF-001_20170601_MWRF-001_724&sfvc4enews=42&cl=article_1&utm_rid=CPG05000006522536&utm_campaign=11360&utm_medium=email&elq2=ceb10e54a8c5428886d074fe32a536dc accessed 9 June 2017

Chapter 3

The future 5th generation (5G) of mobile networks: Challenges and opportunities of an impelling scenario

3.1 General considerations on mobile telecommunication networks

Throughout this chapter, the focus is going to move from RF-MEMS technology to plausible ways of its exploitation into a market scenario as wide and articulated as that of mobile telecommunications. In the previous chapter, the increasing confirmation of RF-MEMS technology in the current market of the 4th Generation Long Term Evolution (4G-LTE) standard smartphones was sketched. From now on, the scope of the discussion will be around the intrinsic potential of RF passives in MEMS technology for the future 5th Generation (5G) of mobile networks. Such a frame of reference is quite complex and dense with ramifications. In light of these considerations, the playground of mobile communications is going to be debated according to an upward/downward cross-reversed approach. If, on one hand, the technical complexity of the analysed items will increase, on the other hand, the level of abstraction will decrease, gliding from a simplified conceptualization of the mobile network as a whole, to the breakdown of some relevant portions, or tiles, composing the mosaic. The envisaged benefit of such an approach is twofold. First, it will make the reader more confident in strolling up and down the field of 5G. Then, it will help to consolidate multiple links between the high-level conceptual reference plane of mobile network and the low-level complexity frame of RF passives, which are comparable, in this circumstance, to a bolt in a skyscraper. Complying with this strategy, the working principles of the network cellular coverage are going to be recalled first.

The fundamental concept of cellular networks is to divide the space covered by the service in limited areas, addressed as *cells*. Each cell is served by a base station, i.e. transmitting and receiving antenna, which is connected (wired) to the central network, managing all the other base stations and the data/information flows across

doi:10.1088/978-0-7503-1545-6ch3

them. The schematic reported in figure 3.1 summarizes how the setting of a cellular network typically appears.

In the example shown in figure 3.1, the transmitting mobile phone (labelled 'S'—speaker) is located in cell A. The signal is received by the cell A base station. Through interrogation of all the other cells composing the network, the receiving mobile phone (labelled 'L'—listener) is located in cell B. Therefore, the cell B base station delivers data to that specific handset. If the person listening to the conversation moves stepping from cell B to cell C, his/her mobile phone is unlinked from the base station of cell B and linked to that of cell C, in a totally transparent way, both to the speaker and listener [1–2]. In spite of the remarkable evolution that telecommunication standards underwent across the timespan of the last four decades, the concept sketched in figure 3.1 is still valid today.

Moving from the network to the mobile phone level of abstraction, it is appropriate to briefly recall the handsets' architecture. The level of detail can be freely set on any plane, however exerting significant impact on the complexity of the scheme and, therefore, on the ease of understanding. For the purposes of this chapter, a high-level block diagram of the smartphone is preferred, splitting its parts based upon the implemented functionalities [3], as reported in figure 3.2.

The RF signal is received from the antenna and treated by the RF transceiver, whose functionality and detailed architecture were previously discussed in this work. The transceiver provides the filtered and amplified baseband signal to the Analogue to Digital Converter (ADC) block. The latter, when dealing with signals to be transmitted, works as a Digital to Analogue Converter (DAC), performing the opposite function. The ADC/DAC block is controlled by the Central Processing

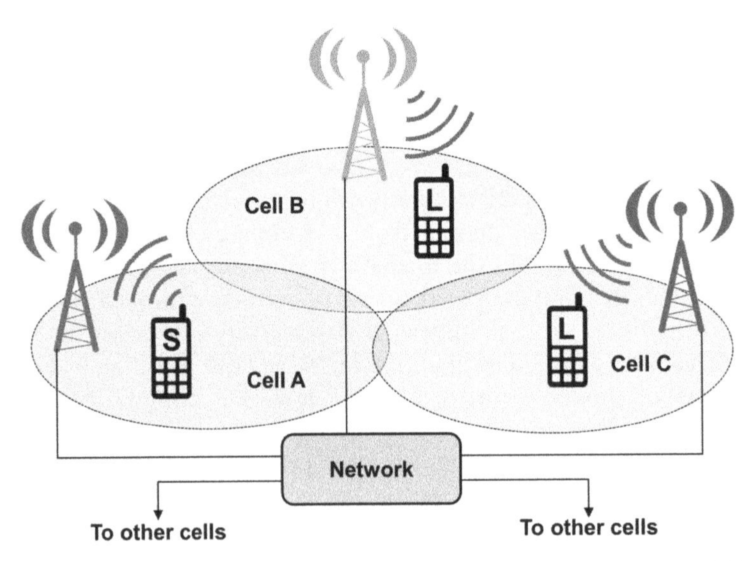

Figure 3.1. Schematic of a typical arrangement of a cellular mobile network. The area is divided into cells, each of which is covered by a base station. The label 'S' identifies the speaker (mobile phone transmitting data), while the label 'L' identifies the listener (mobile phone receiving data).

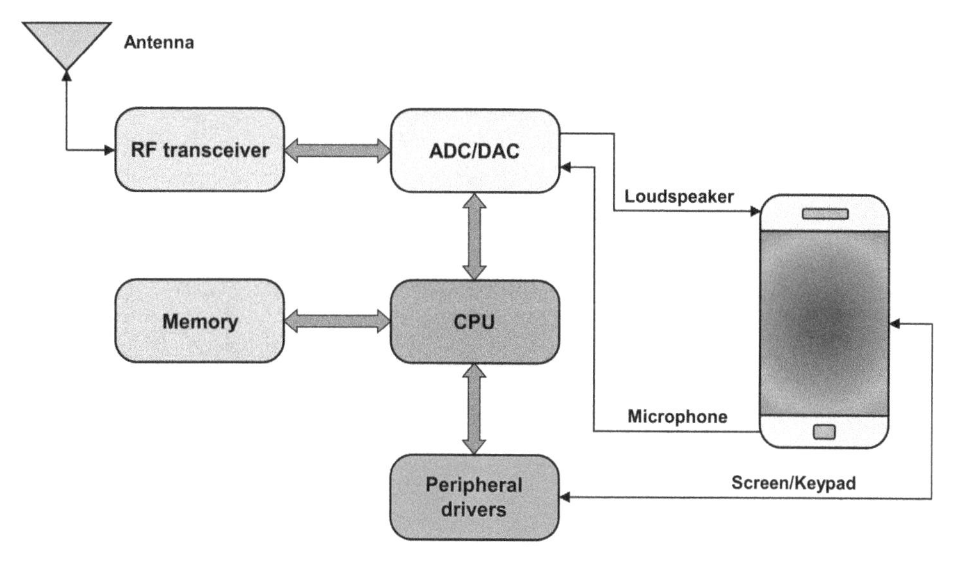

Figure 3.2. High-level block diagram of a modern smartphone. The block splitting is based on the functionalities implemented by each of them.

Unit (CPU), which performs other treatment functions on the received signal in the digital domain, before driving it into the loudspeaker. Similarly, the voice collected by the microphone follows the opposite path through the ADC/DAC block (digital to analogue conversion) and the RF transceiver (filtering, amplification and frequency up conversion), before being transmitted by the antenna. The CPU also controls the memory unit and the peripheral drivers block, thus driving all the other on-board devices, like touchscreen (input/output device) and sensors (gyroscopes, accelerometers, cameras, and so on). The schematic reported in figure 3.2 is notably simplified, as critical functions such as the encoding/decoding of signals for instance, is not mentioned. However, it provides a sound and easy-to-catch insight of modern mobile phones' architecture.

Before concluding this introductory section on mobile applications, the level of abstraction is going to be taken for a while to the highest level. The target is to complete the overview provided up to here, with a brief insight into the overall socio-economic environment related to mobile market applications, thus featuring all the profit and non-profit stakeholders. To this purpose, the schematic representation in figure 3.3 is provided.

The main pillars of the mobile applications ecosystem are embodied by the following entities: (a) networks and infrastructures; (b) mobile phones; (c) end users. As a matter of fact, very simply, end users buy and use mobile phones, which need proper infrastructures to work (horizontal arrows in figure 3.3). Above the pillars stand network operators, which interact directly with all the former entities (vertical arrows in figure 3.3). Network operators provide diverse services to end users, also including mobile phones. On the other hand, they manage, and very often own, mobile infrastructures. In the lower part of the scheme are mobile phone makers and infrastructure makers/providers. Mobile phone makers interact directly

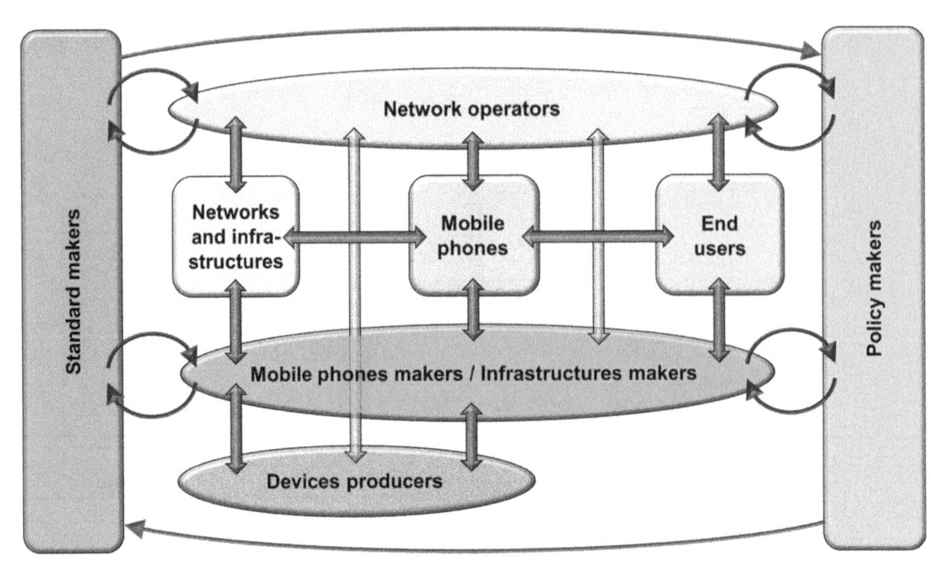

Figure 3.3. Schematization of the socio-economic environment featuring the most relevant categories of stakeholders, spinning around mobile applications.

with end users, as they develop products complying with the needs and expectations of customers. On the other hand, infrastructure makers develop and provide network terrestrial infrastructures. At the very bottom stand device producers, often called Original Equipment Manufacturers (OEMs). Their role is to develop and sell components both to mobile phone and infrastructure makers, necessary in the assembly of their respective products. Network operators also interact with mobile phone/infrastructure makers as well as product makers (lighter vertical arrows in figure 3.3). This is because operators guide/influence the specifications and requirements mobile handsets and infrastructures have to reach.

The mobile ecosystem is populated by other two critical stakeholders, namely standard and policy makers. The former act on the definition that mobile communication standards have to comply, while the latter define the regulations to be applied to mobile services, in accordance to existing national and international laws, e.g. concerning frequency band allocations, electromagnetic compatibility, and so on. Just to mention a few of them, the 3rd Generation Partnership Project (3GPP) and the International Telecommunication Union (ITU), are examples of such classes of stakeholders. Standard and policy makers have mutual interaction with network operators and mobile phones/infrastructure makers, as well as between themselves. Their role is crucial both during the preliminary definition of standard specifications, as well as after the release and deployment of a certain generation of mobile services, for control/management purposes.

The latter description concludes the unwinding of the introductory elements meant to ease understanding of the next few sections. In the following, the evolution of telecommunication standards will be discussed, from the early days of mobile services, approaching in a smooth fashion the core topic of 5G.

3.2 Evolution of mobile standards and services

The focus is now going to be shifted to the progression of mobile telecommunication standards, over the last four decades. The most relevant achievements, challenges and limitations of mobile services, starting from the first generation (1G) to the current 4G-LTE, will be listed, thus building a proper background for unfolding the scenario of targets and advancements to be reached by the future 5G standard. As a matter of fact, already starting from 1G, deployed across the turning point between the 1970s and the 1980s, the technology advancement was unambiguous with respect to the first mobile device (i.e. a radio signal strength testing system), dated 1924 and shown in figure 3.4.

3.2.1 The 1st generation—1G

The first commercial, fully automated cellular network generation (1G) for voice transmission was launched in Japan in 1979, providing coverage of the full metropolitan area of Tokyo (more than 20 million inhabitants). The 1G network expanded to cover the whole population of Japan within 5 years from the first deployment. The second launch of 1G networks was performed in Denmark, Finland, Norway and Sweden in 1981. Subsequent 1G networks were launched in

Figure 3.4. First *portable* device for mobile radio communications (i.e. a radio signal strength testing system), dated back to 1924. Reused with permission of Nokia Corporation.

United Kingdom, Mexico, Canada and United States in the early 1980s [4]. The main mobile phone cell systems standards spreading under the umbrella of 1G were the Advanced Mobile Phone System (AMPS), Nordic Mobile Telephone (NMT) and Total Access Communication System (TACS) [5].

1G established the fundamental pillars of mobile communications that, albeit undergoing significant evolution in terms of complexity, were to be exploited in the subsequent generations of cellular services. The general setting of 1G networks is the one previously depicted in figure 3.1. The transmission of voice was analogue and taking place over a licensed spectrum, i.e. a cleared spectrum exclusively allocated for mobile communications. The terrestrial infrastructure, i.e. base stations and network as reported in figure 3.1, was deployed by mobile operators [6]. 1G established seamless access to mobile handsets by means of transparent backhaul network, as well as full integration with Public Switched Telephone Network (PSTN). Moreover, since one channel was allocated for just a single voice signal, 1G introduced the concept of frequency reuse, as illustrated in figure 3.5.

Adjacent base stations were using different frequencies, in order to avoid interference. Nonetheless, in the span of a few cells, the same frequency was reused (cells with the same colour in figure 3.5). As mentioned before, 1G was transmitting one voice signal per channel. Therefore, separation between different channels, i.e. among different voice conversations, was achieved by setting them apart over frequency, thus implementing the so-called Frequency Division Multiple Access (FDMA), schematically shown in figure 3.6.

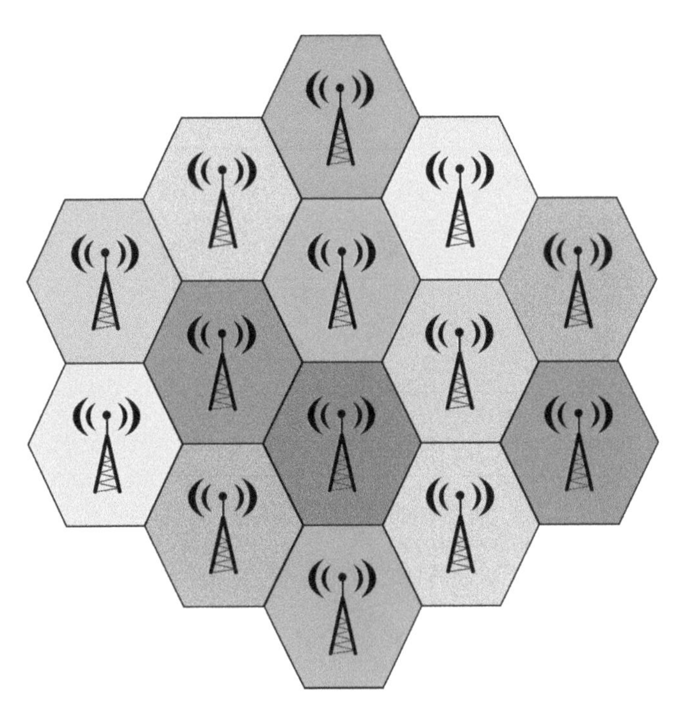

Figure 3.5. Frequency reuse across non-adjacent cells, introduced in the 1G to reduce frequency occupation without creating interferences. The same colour indicates the same frequency.

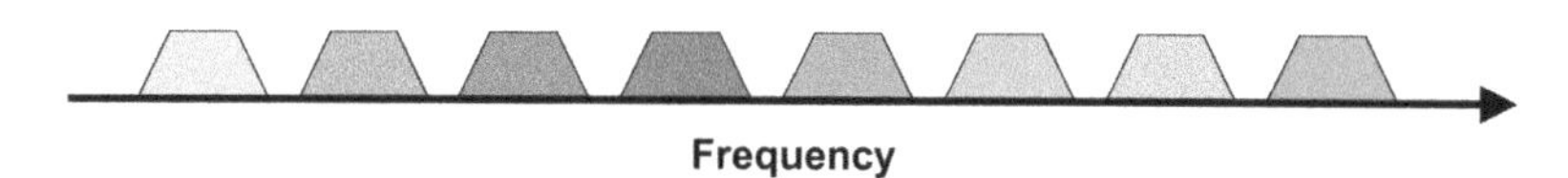

Figure 3.6. Schematic representation of the Frequency Division Multiple Access (FDMA) between different voice channels adopted by 1G.

Each channel was allocated 30 kHz band and needed certain clearance in terms of frequency distance from other channels, in order to avoid interference. In light of the aforementioned characteristics, 1G introduced significant innovation, but was also characterized by several limitations. In the first place, the FDMA in figure 3.6 and the allocation of one signal per channel was definitely non-optimal in terms of spectrum usage efficiency. Hard handoffs in switching channel frequency, when passing from one base station to another, were also causing dropped calls quite frequently. In addition, 1G was not secure at all, as voice calls were broadcasted with no encryption [7]. Also an important point to note is that analogue handsets were bulky (with limited potential for downscaling), power inefficient and expensive.

3.2.2 The 2nd generation—2G

The second generation of mobile networks (2G), released in the early 1990s, had the appropriate characteristics to trigger the spread of mobile services to large numbers of people. In contrast to the 1G standard, 2G was exclusively dealing with digital signals. Stepping into the digital world enabled two significant advancements in terms of transmission efficiency. On one hand, digital voice signals could be encoded, thus significantly compressing the amount of data to be broadcasted. Just to report a few numbers, an uncompressed voice signal required 64 kb (kilobit) per second, while after encoding the necessary bitrate was reduced to 8 kb per second [6]. On the other hand, with 2G it was possible to transmit more than one voice signal over the same channel, overcoming the 1G limitation of one channel per one signal. This was possible by splitting different voice signals into packets of data and broadcasting them over a unique channel, according to the so-called Time Division Multiple Access (TDMA). In particular, the Digital Advanced Mobile Phone System (D-AMPS), mainly used in North America, was ensuring the transmission of three different voice signals (talking one at a time) over the same 30 kHz wide channel. However, the actual wide spread of 2G mobile services took place with the introduction of the Global Systems for Mobile communications (GSM), which allowed transmission of eight different voice signals over the same 200 kHz wide channel [6]. Also, GSM allowed transmission of the first non-voice data, by means of Short Message Service (SMS) and e-mail exchange. The 2G D-AMPS and, in particular, the GSM, were both based on TDMA, so were still suffering as 1G from large frequency gaps needed between channels to avoid interference, as well as from frequently dropped calls in hard handoffs, at the edge from one cell and another.

What radically improved the traffic volume capacity in the evolved protocols released under 2G, posing also as the basis of 3G, was the development of the Code Division Multiple Access (CDMA) approach [8]. The rationale underlying CDMA is shown schematically in figure 3.7.

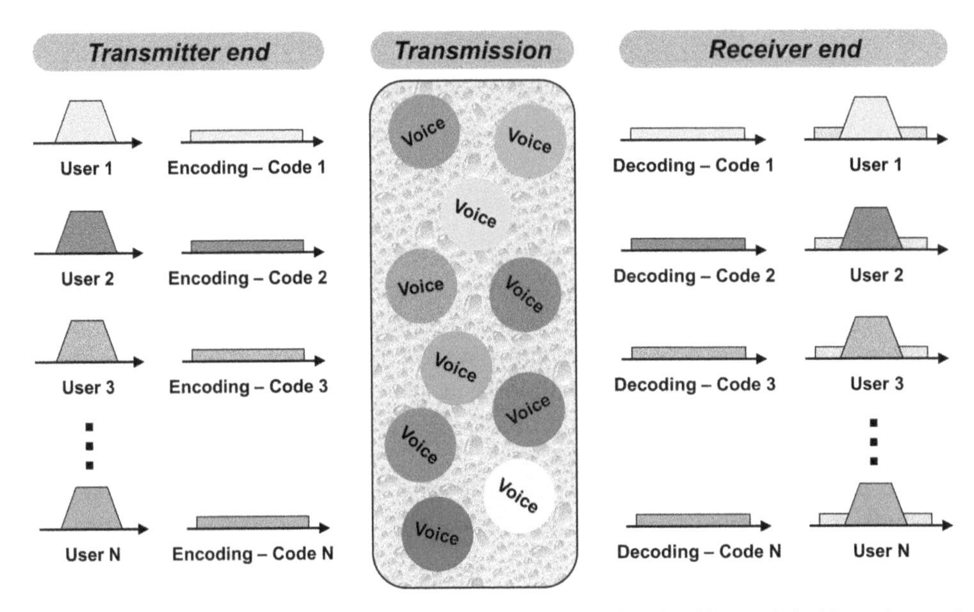

Figure 3.7. Schematization of the CDMA transmission. Each user voice signal is encoded with a unique code and transmitted together with other voice signals. When received on the listener side, the voice signal is reconstructed (decoded) by using the same code. Signals related to other users appear as noise (below noise threshold).

On the transmitter end, the voice signal of each user is encoded with a unique code, and transmitted together with another voice signal over 1.25 MHz wide channels. At the receiver end, the voice signal is reconstructed (decoded) using the same code originally exploited for encoding. This makes it possible to deliver the proper voice signal of the speaker to the proper listener, with all the other signals appearing as background noise. The CDMA exhibited several advantages in comparison to previous TDMA. First, voice capacity was significantly increased, as the whole frequency spectrum was finally exploited across its extent. CDMA also improved the security of communications, as encoding made eavesdropping much more challenging. Furthermore, efficiency improvements in the transmission of data led also to an increase of battery life in mobile handsets, which was definitely a plus, especially in respect of 1G.

Nonetheless, a few technical challenges had to be solved before making CDMA successful for commercialization. Among them, the near/far power challenge was one of the most relevant. When the user was close to the base station, the tower was overpowering the uplink channel, thus significantly reducing the band available to other users located farther away from the cell infrastructure. This issue was circumvented by introducing continuous transmitted power control on the handset side, depending on the strength of the link signal to the base station. In addition, interference between adjacent base stations when the user was at the boundary between cells was also a relevant issue. This challenge was addressed by encompassing the possibility for the handset to communicate simultaneously with both towers, thus smoothing the handoff process. After addressing the mentioned problems,

CDMA was released commercially in the mid-1990s, under the cdmaOne (IS-95) standard.

Despite the fact that CDMA was not already employed at a commercial level, 2G evolved further based upon TDMA, by introducing protocols with enhanced transmission data rates and with more pronounced use of packet switching besides traditional circuit switching. This brought advanced (sub-)generations of 2G, commonly referred to as 2.5G and 2.75G [5]. In particular, in 2.5G the General Packet Radio Services (GPRS) was launched (data rates up to 115 kbps), while in 2.75G the Enhanced Data Rate for GSM Evolution (EDGE) was also employed (data rates up to 384 kbps).

3.2.3 The 3rd generation—3G

Capitalising on the important advancements introduced by 2G, the 3G was launched in 2000. Before listing the achievements at service and implementation level brought by 3G, a couple of brief, general considerations are necessary. At the turning of the millennium, people were getting used to the Internet and having broadband access, both in the office and home. This scenario was also driving the increasing demand for Internet access via mobile phones. However, handsets' technologies were taking significant steps forward, introducing relevant innovations, not only in RF systems and transmission data capabilities, but also in relation to other components integrated in the device, such as cameras and screens with improved characteristics. Also, after the widespread distribution of 2G, more and more people had a mobile subscription, and most of them were expecting more capacity in terms of data access on the move. Therefore, it is straightforward that 3G had to deal with all of the mentioned trends and demands.

From the viewpoint of voice transmission, 3G consolidated the exploitation of the CDMA technology by means of the CDMA2000 standard (channel width of 1.25 MHz). Nonetheless, while in 2G, transmission of data (for SMS, e-mail and news services) was travelling together with the voice signal, whereas 3G introduced a paired channel entirely dedicated to non-voice services. In particular, the Evolution-Data Optimized (EV-DO) introduced a data-only 1.25 MHz optimized channel for Internet, multimedia and navigation services. This solution ensured access to richer content and provided the main pillars of mobile broadband. To this regard, the EV-DO introduced the concept of adaptive modulation, which means using a higher order of modulation when the signal strength and quality are good, such as when the handset is close to the base station. Increasing the modulation order means improving the spectral efficiency in terms of bits per second per Hertz (bps/Hz) and, in turn, the amount of transmitted/received data, keeping the bandwidth fixed. Moreover, the management of transmission according to time slots typical of EV-DO, enabled the implementation of opportunistic scheduling concepts. In other words, channel use was optimized, scheduling transmission of data when users established a good link to the base station in terms of signal strength.

The combination of CDMA2000 and EV-DO brought 3G to the further evolution of Wideband Code Division Multiple Access (WCDMA), Universal

Mobile Telecommunications System (UMTS) and High Speed Uplink/Downlink Packet Access (HSUPA/HSDPA). In particular, the concurrence of the just-mentioned protocols brought an increase of data rate in the range from 5 Mbps to 30 Mbps, commonly called 3.5G [5]. In particular, the Evolved High Speed Packet Access (HSPA+) and EV-DO Rev. B (Revision B) ventured into the fields of higher order modulation and Carrier Aggregation (CA). Concerning the former, Quadrature Amplitude Modulation (QAM) with 64 possible combinations (64-QAM) was explored, thus transmitting 6 bits per symbol, which led to an increase of about 50% in terms of bps/Hz. On the other hand, CA enabled one to extend the spectrum by combining diverse carrier signals and putting together different frequency bands and channels. These technologies, among other benefits, enabled reduced operator cost for data services and built a platform triggering continual enhancement of the functionalities provided to the final users. Eventually, the further released standards, like Worldwide Interoperability for Microwave Access (WiMAX) and Ultra Mobile Broadband (UMB) led to a sort of seamless transition between 3G and 4G, across the so-called 3.75G.

3.2.4 The 4th generation—4G

The scenario depicted in the previous pages elucidated the trends that mobile telecommunications have been seamlessly following for several years, stepping from one generation to the next, as well as within the same generation, introducing continual innovations and improvements. Given such a landscape, 4G Long Term Evolution (4G-LTE) is not an exception. Integration of more services (download, browse, stream, etc) under the umbrella of 4G, poses non-reversible demands in terms of faster and more efficient data transfer capacity. In addition, another self-sustaining mechanism triggering the former demands should also be borne in mind. When mobile services supported by modern smartphones combine with good network performance and desirable content, the flywheel effect shown in figure 3.8 is established, leading to further increase of data consumed by end users [9].

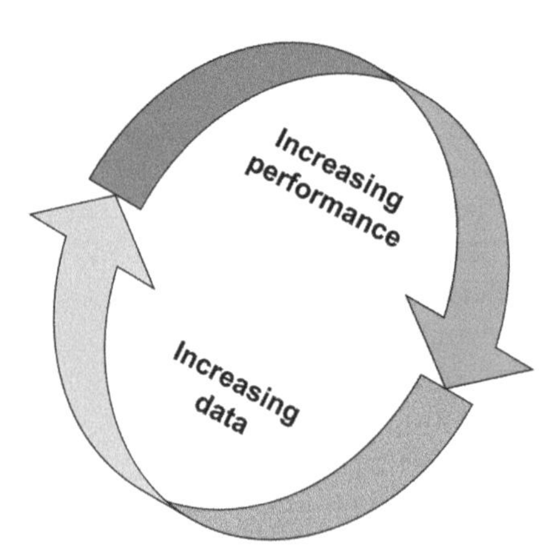

Figure 3.8. Flywheel effect triggering more and more data demand from the end user side.

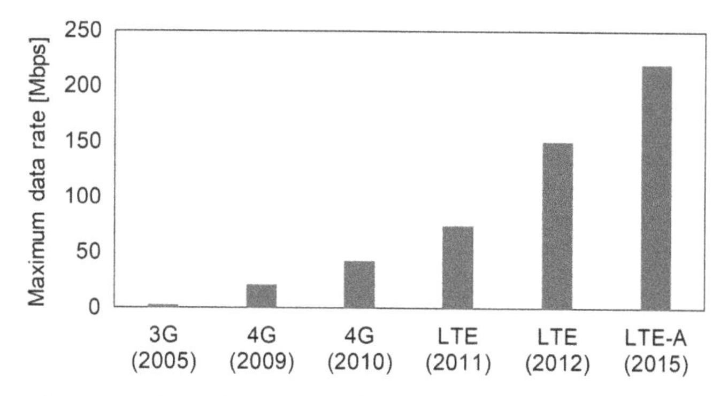

Figure 3.9. Comparison of maximum data rate capacity, in terms of Mbps, from 3G to early releases of 4G-LTE, also including LTE Advanced (LTE-A). The year of release is indicated per each generation.

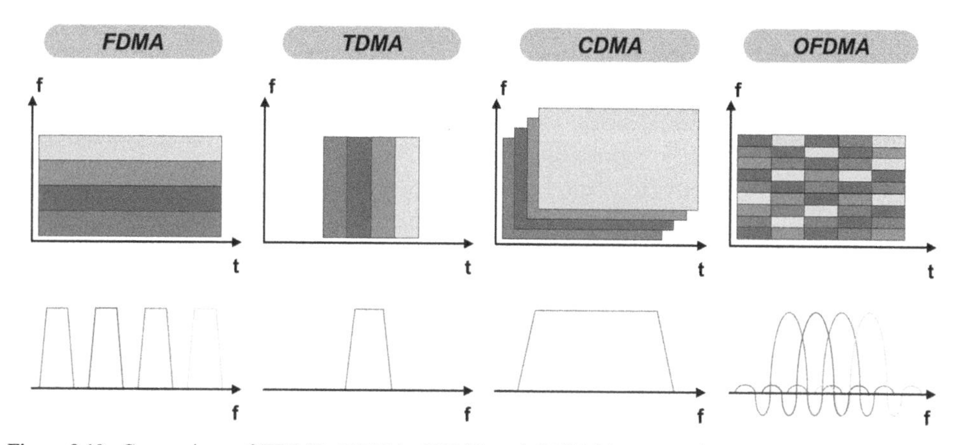

Figure 3.10. Comparison of FDMA, TDMA, CDMA and OFDMA, versus time and frequency. OFDMA exploits numerous subcarriers to build a band that is wider in comparison to other multiple access strategies.

One can clearly understand the relevance of the above-mentioned trend simply by looking at how maximum data rates (in terms of Mbps) increased when stepping from 3G to 4G, and then, within 4G-LTE, from one release to the next. For the benefit of this purpose, the plot in figure 3.9 shows the maximum data rates, also reporting the year of release [10].

Bearing in mind the scenario just depicted, 4G addresses the increasing demands, acting at different levels. From a technical point of view, 4G enables flexible support of channels up to 20 MHz by exploiting the Orthogonal Frequency Division Multiple Access (OFDMA) [11]. The latter concept can be easily explained if compared to the other multiple access techniques, previously discussed when listing former generations of mobile communication standards. To this purpose, figure 3.10 shows the time and frequency characteristics of FDMA, TDMA, CDMA and OFDMA. Each colour corresponds to the band allocated to a different user.

In the FDMA scenario, the band occupied by each user is a horizontal strip on the frequency versus time plot (top of figure 3.10), corresponding to separated bands on the frequency axis (bottom of figure 3.10). In the TDMA case, instead, each user has a vertical strip on the frequency versus time plot, yielding a unique frequency band on the bottom plot (as its occupation is split over time). However, CDMA exploits unique encoding/decoding to transmit, identify and reconstruct the signal of a certain user (see figure 3.7). Therefore, data of all users are broadcasted together over a unique frequency band (wider than in FDMA and TDMA). Instead, a radically different approach is pursued by OFDMA. The spectrum is fractioned down in a variable number of narrow bands (15 kHz wide in LTE) built around multiple sub-carriers. The latter are orthogonal, meaning that the centre frequency of each narrow band intercepts nulls of adjacent subcarriers [11]. Each narrow band carries a limited amount of information, albeit the whole throughput is given by the sum of all the employed subcarriers, leading to a rather wide effective band. The significant advantage of OFDMA is that the width of the band associated to each user can be widened or narrowed depending on the amount of data to be transferred at a given time, thus making part of it available to other users when not utilized.

Returning to the technical innovations introduced by 4G-LTE, co-existence of paired and unpaired spectrum for Downlink (DL) and Uplink (UL) is also supported. In more detail, both Frequency Division Duplex (FDD) and Time Division Duplex (TDD) are possible. In FDD mode, the DL and UL spectra are paired and lie on different frequency bands. In contrast, in TDD mode, the DL and UL spectra are unpaired, lie on the same frequency band and are assigned different timeslots [12]. Both FDD and TDD have pros and cons, also including space coverage, as reported in figure 3.11.

In particular, FDD enables coverage of large areas, while TDD makes asymmetry of DL and UL possible, thus extending the DL capacity when needed. However, TDD is able to cover reduced areas, mainly because the UL device power is used part of the time in TDD mode, but continuously in FDD mode.

To conclude the overview of technical trends followed by 4G-LTE, another two aspects must be mentioned. On one hand, further advancement of Carrier Aggregation (CA) techniques, previously introduced when discussing 3G, is leading and will lead to wider data pipes and, in turn, increased data rates. Also relevant, 4G-LTE capitalises on techniques aimed to sustain spatially separated multiple transmission paths. To this regard, Multiple Input Multiple Output (MIMO) techniques are massively pursued to enable multi-path broadcasting [13–14]. They consist in having multiple antennas, both on the transmitter and receiver end, in order to transmit/receive in parallel through distinct and frequency diverse channels. The schematic in figure 3.12 shows the case of 2×2 MIMO, in which two antennas are featured at both ends.

The advantages introduced by employing MIMO solutions are multiple. As first instance, the overall data rate can be significantly increased. Moreover, interferences and/or non-optimal transmission/reception, due to the presence of obstacles or weak signals, can be significantly mitigated by exploiting different frequencies. In addition, handoffs at the cells' boundary could be made even smoother. Of course, these improvements can be further enhanced acting on the solution complexity. This is the reason why massive MIMO solutions are being investigated, featuring more than just two antennas for each end [9].

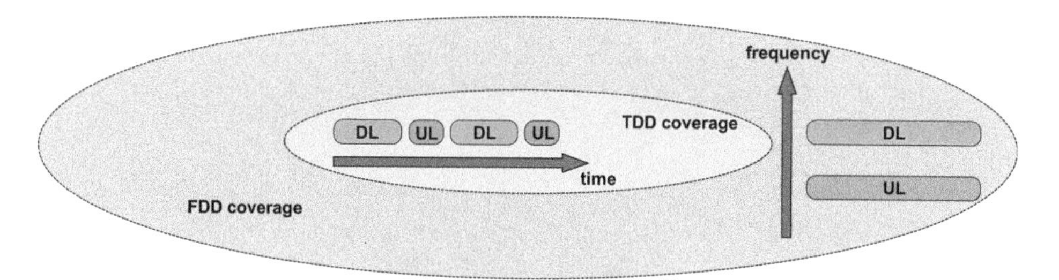

Figure 3.11. Schematic of Frequency Division Duplex (FDD) and Time Division Duplex (TDD) related to Downlink (DL) and Uplink (UL), also including space coverage.

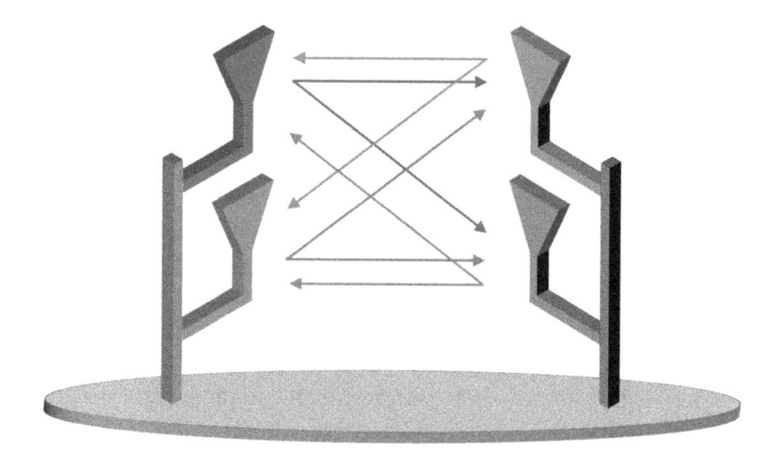

Figure 3.12. Schematic example of 2x2 Multiple Input Multiple Output (MIMO) solution, featuring two antennas both on the transmitter and receiver end.

For the sake of completeness, it must be mentioned that 4G-LTE is pushing forward significant innovation in terms of infrastructure and network settings. By summarizing in just three keywords the trends followed at high-level, it can be stated that 4G-LTE pursues *virtualization*, *densification* and *diversification* of the network.

As demand and configuration requirements for terrestrial infrastructures are increasing in terms of flexibility and adaptability, the overall architecture is following a flattening trend that complies with the latter requests and simplifies, at the same time, the hardware architecture. This important target is addressed by virtualization of network functions, together with the increasing adoption of Internet-based protocols to manage and reconfigure such functionalities [15–16].

Recalling once again the increasing trends in data throughput and connectivity followed by 4G-LTE, network densification and diversification are other key-points, intimately linked together, as well as correlated to virtualization. It is straightforward that improved services can be made available only if the network is deployed in a capillary fashion (i.e. network densification) [17–19]. However, this target cannot be reached by simply increasing the number of base stations available, but rather by pursuing diversification of the network (i.e. heterogeneous network) at the same

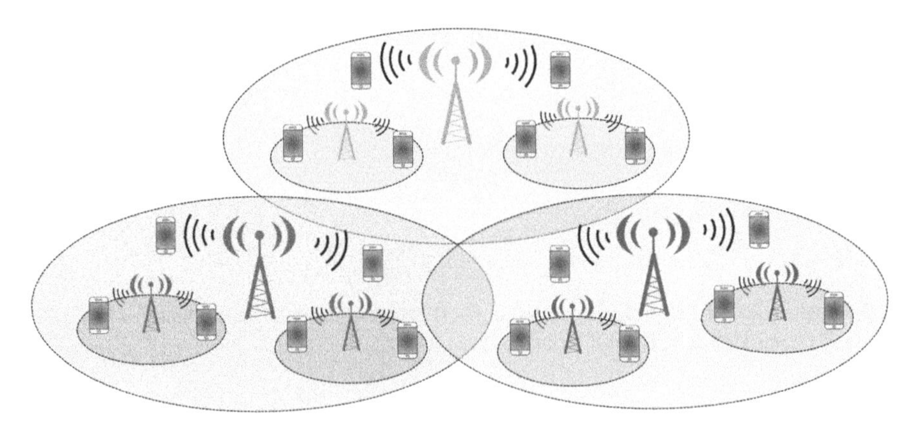

Figure 3.13. Schematic representation of the network diversification concept, including cells with different area coverage.

time. In practical terms, this means extending the backhaul with diversified radio towers, in terms of delivered power and covered area, and being able to provide higher throughputs and pronounced connectivity where more is needed (e.g. metropolitan versus rural areas) [20–23]. Network diversification led to the introduction of concepts like Macro, Micro, Pico and Femtocells; addressing a variety of radio towers with scaled power, coverage and data volume capacities, realizing the hierarchical arrangement in figure 3.13.

In order to conclude this section on 4G-LTE, the current state of its evolution is going to be reported in brief, by discussing some of the most relevant advancements at the technical level. Starting from the CA technique, introduced for the first time in Release 10 (in 2011) with two DL carriers; in the current Release 14 it features up to 5 DL and 3 UL carriers [9]. Concerning the order of modulation, 64-QAM in DL and 16-QAM in UL were initially introduced. In the following releases, they were increased to 256-QAM in DL and 64-QAM in UL. The current Release 14 is going to reach 256-QAM in UL as well. Furthermore, dealing with multi-paths, 4×4 MIMO has already been implemented in the most advanced networks where, in combination with CA with 3 DL components and 256-QAM, peak data rates quite close to 1 Gbps were demonstrated.

3.2.5 Wrap up of mobile standards from 1G to 4G

This conclusive part aims to collect the most salient features that were mentioned in previous pages, in order to enable an easier and more direct understanding of the differences and advancements, in the seamless evolution that brought, across more than four decades, 1G to the most recent 4G-LTE releases.

In the first place, it is interesting to report on the lifecycle of mobile generations across the whole timespan including research and technology development, standard definition, deployment, commercial spread and, finally, decrease and fading-out from the market scenario. To this regard, an effective interpretation is provided in [24] and graphically sketched in figure 3.14.

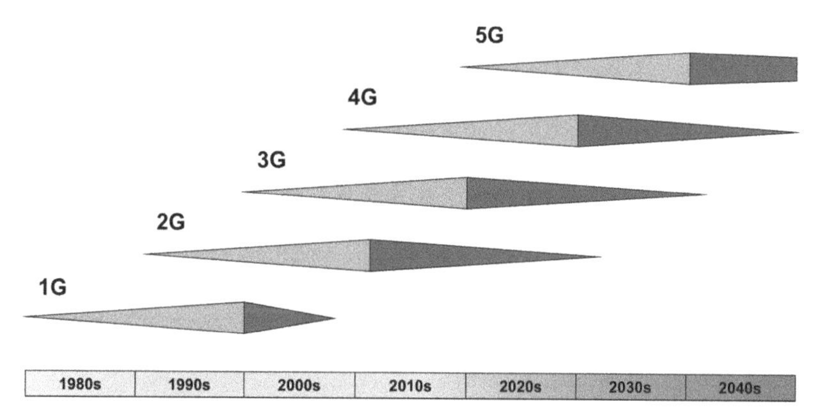

Figure 3.14. Expansion and contraction phase over time of mobile generations from 1G to 5G.

The scheme reports a kite-like bar over time per each generation. The widening (green) part refers to the timespan from the beginning of development to the maximum spread in terms of market volumes and number of users. On the other hand, the narrowing (red) part provides a time indication concerning the fading-out of a certain generation, until its complete obsolescence and disappearance from the market. Looking at the graph in figure 3.14 a few considerations emerge quite evidently. First, the time necessary for a certain technology to reach its maximum commercial spread, starting from early experimentation and deployments, is as long as two decades. Also surprisingly, albeit we are fully immersed in the 4G-LTE and focused on future 5G, the widest worldwide market of 2G started declining just a few years ago, while that of 3G is still to come.

Finally, a comprehensive summary [5] of the main technical characteristics discussed up to now is reported in table 3.1, thus concluding the section.

3.3 The future 5th generation—5G

The ample overview provided on previous and current generations of mobile communications standards, helps one to identify the main underlying trends that, despite the interchange between technical solutions and technologies, will linger on in being pursued. From this perspective, it is straightforward that the future 5th generation (5G) will be required to provide more data transfer capacity and increased coverage. Nevertheless, in order to aggregate a more pragmatic vision around the needs that 5G will have to address, one should wonder about the metamorphosis the Internet has been going through in recent years.

We are living in a world where the surrounding environment and objects supporting us in daily life are seamlessly becoming more adaptive and interconnected. Services are broader, more capillary, linked together and easy to access. Experiences are tailored to our preferences and specific (as well as special) needs. The attribute *smart* is now frequently associated with diverse application fields of science and technology, giving rise to familiar keywords like Smart Cities, Smart Homes, Smart Objects, etc. The Internet of Things (IoT) paradigm portrays an ongoing technology development path through which any object and environment

Table 3.1. Summary of the most relevant technical characteristics previously discussed for mobile generations from 1G to 4G-LTE.

Generation	Standard/Technology	Data rate	Band	Bandwidth	Switching	Services
1G	AMPS; TACS; NMT – FDMA	2.4 kbps	800 MHz	30 kHz	Circuit	Voice
2G	GSM – TDMA CDMA	10 kbps	850 MHz; 900 MHz;	200 kHz 1.25 MHz	Circuit	Voice/Data
2.5G	GPRS – TDMA EDGE – TDMA	50 kbps 200 kbps	1.8 GHz; 1.9 GHz	200 kHz 200 kHz	Circuit/Packet	
3G	WCDMA; UMTS CDMA2000	384 kbps 384 kbps	800 MHz; 850 MHz;	5 MHz 1.25 MHz	Circuit/Packet Circuit/Packet	Voice/Data/Video calling
3.5G	HSUPA/HSDPA EV-DO	5–30 Mbps 5–30 Mbps	900 MHz; 1.8 GHz; 1.9 GHz; 2.1 GHz	5 MHz 1.25 MHz	Packet Packet	
3.75G	LTE – OFDMA/SC-FDMA[1] Fixed WiMAX – SOFDMA[2]	100–200 Mbps 100–200 Mbps	1.8 GHz; 2.6 GHz 3.5 GHz; 5.8 GHz (initially)	From 1.4 MHz to 20 MHz 3.5 MHz, 7 MHz in 3.5 GHz band; 10 MHz in 5.8 GHz	Packet	Voice/Data/Video calling/Online gaming/High definition television
4G	LTE-A[3] – OFDMA/SC-FDMA[1] Mobile WiMAX – SOFDMA[2]	DL 3 GbpsUL 1.5 Gbps 100-200 Mbps	1.8 GHz; 2.6 GHz 2.3 GHz; 2.5 GHz; 3.5 GHz (initially)	From 1.4 MHz to 20 MHz 3.5 MHz; 5 MHz; 7 MHz; 8.75 MHz; 10 MHz	Packet	Voice/Data/Video calling/Online gaming/High definition television

[1] Single Carrier Frequency Division Multiple Access
[2] Scalable Orthogonal Frequency Division Multiple Access
[3] Advanced Long Term Evolution

belonging to our daily life experience earns its own identity in the digital world by means of the Internet. The IoT is based on smart objects/environments with one or more of the following functionalities [25–27]:

- Self-awareness, i.e. identification, localization and self-diagnosis of the object/environment;
- Interaction with the surrounding environment, i.e. data acquisition (sensing and metering) and actuation;
- Data elaboration, both basic (primitive data aggregation) and advanced (statistics, forecasts).

Regardless of the set of functionalities, all smart objects/environments must feature data transfer capabilities (wired or wireless) in order to be networked and framed within the IoT. Despite the fact that the fields of application are widely diverse, it is possible to identify macro-areas for which IoT represents the common denominator, as follows: **(1)** Smart City/Environment; **(2)** Smart Home; **(3)** Smart Metering/Smart Grid; **(4)** Smart Building; **(5)** eHealth; **(6)** Smart Logistics; **(7)** Smart Factory; **(8)** Smart Asset Management; **(9)** Smart Agriculture; **(10)** Smart Car.

On a higher-level reference plane, the concept of IoT is somehow overtaken, or, better, further generalized by the paradigm of the Internet of Everything (IoE) [28], meaning that connectivity must be provided also without the perception of a physical object (i.e. a *thing*) as intermediate mean. In other words, the IoE encompasses application domains such as Virtual Reality (VR), Augmented Reality (AR), remote (cloud) computing, Machine to Machine (M2M) communication and the Tactile Internet [29].

As the IoT/IoE weigh on the shoulders of 5G, it is clear that the demand in terms of data throughput will be massive. Many predictions appeal for an enhancement in 5G transmission capacity as big as 1000 times with respect to 4G-LTE, delivering 10 Gbps to each individual user [30]. In addition, latency will need drastic reduction to the millisecond-level. In order to embrace the importance of the latter requirement, one can simply wonder about how low-latency can be crucial for applications like Vehicle to Vehicle (V2V) communications. Finally, yet importantly, with leverage on M2M applications, cloud computing, IoT and so on, more symmetry between DL and UL data transmission capacity will be demanded for 5G standard. In light of these challenges, the main vectors defining the success of 5G have been defined as follows [31–33]:

- **EMBB** (Enhanced Mobile Broadband) is related to the system capacity growth, mentioned above, with the target of 1000 times increase, leading to 10 Gbps peak and a minimum of 10 Mbps per each user. This aspect is also linked to the used spectrum, i.e. below 6 GHz and above 6 GHz 5G New Radio (NR) communications;
- **URLLC** (Ultra-Reliable Low-Latency Communications) is demanded by critical applications, like the aforementioned V2V communications and will aim towards the millisecond latency range;
- **MMTC** (Massive Machine-Type Communications) will lead to a very-large increase of new connections to be supported by 5G services.

The way that these fundamental drivers are pursued cannot be unfolded in unsophisticated terms, as the complexity of the whole frame of reference is significantly pronounced. It is more convenient to decompose the overall scenario into the fundamental challenges and facilitating elements that can support them across the process through which they are distilled in design fundamentals to be complied with, as proposed in [34–35]. The plot of cross-relationships reported in figure 3.15 provides a graphical interpretation of the aforementioned concept.

In order to complete the information shown in figure 3.15, a summary of all the acronyms reported in the scheme is presented in table 3.2.

Pursuing the aforementioned main vectors through the strategic plot shown in figure 3.15 will be unavoidable in pushing forward the vision of the all-communicating world, already undertaken by currently used mobile services. The centrality of such a statement emerges even more clearly if one bears in mind the forecast of 50 billion devices connected to the cloud, expected to be reached by 2020. For the sake of clarity, the six main challenges listed in figure 3.15, from the point of view of network requested performance and characteristics, could be inflected as follows [36]:

1) Data volume increased up to 1000 times with a growth of indoor data traffic up to 70%;
2) Connected devices increased from 10 to 100 times;
3) Typical user data rate increased from 10 to 100 times;
4) Extended battery life up to 10 times for Massive Machine Communication (MMC) devices;
5) E2E latency reduced by 5 times.

The challenge of providing a common connected platform that is able to reach the goals listed will require acting at diverse technical levels, among which the most relevant ones can be identified as reported in the following [36]:

- **Radio links**. New transmission waveforms and new approaches for multiple access control and resource management;
- **Multi-node/multi-antenna transmissions**. Design of multi-antenna transmission and reception technologies, capitalising on massive antenna configurations and enabling advanced inter-node coordination schemes and multi-hop technologies;
- **Network dimension**. Novel approaches for efficient interference management in complex heterogeneous deployments;
- **Spectrum usage**. Operation in extended spectrum band and in new spectrum regimes, to provide a complete system concept able to address specific needs of diverse usage scenarios.

To complete the discussion, a visual representation of how the 5G network could be architected is shown schematically in figure 3.16.

In order to complete the information shown in figure 3.16, a summary of all the acronyms reported in the scheme is presented in table 3.3. As visible, the previously

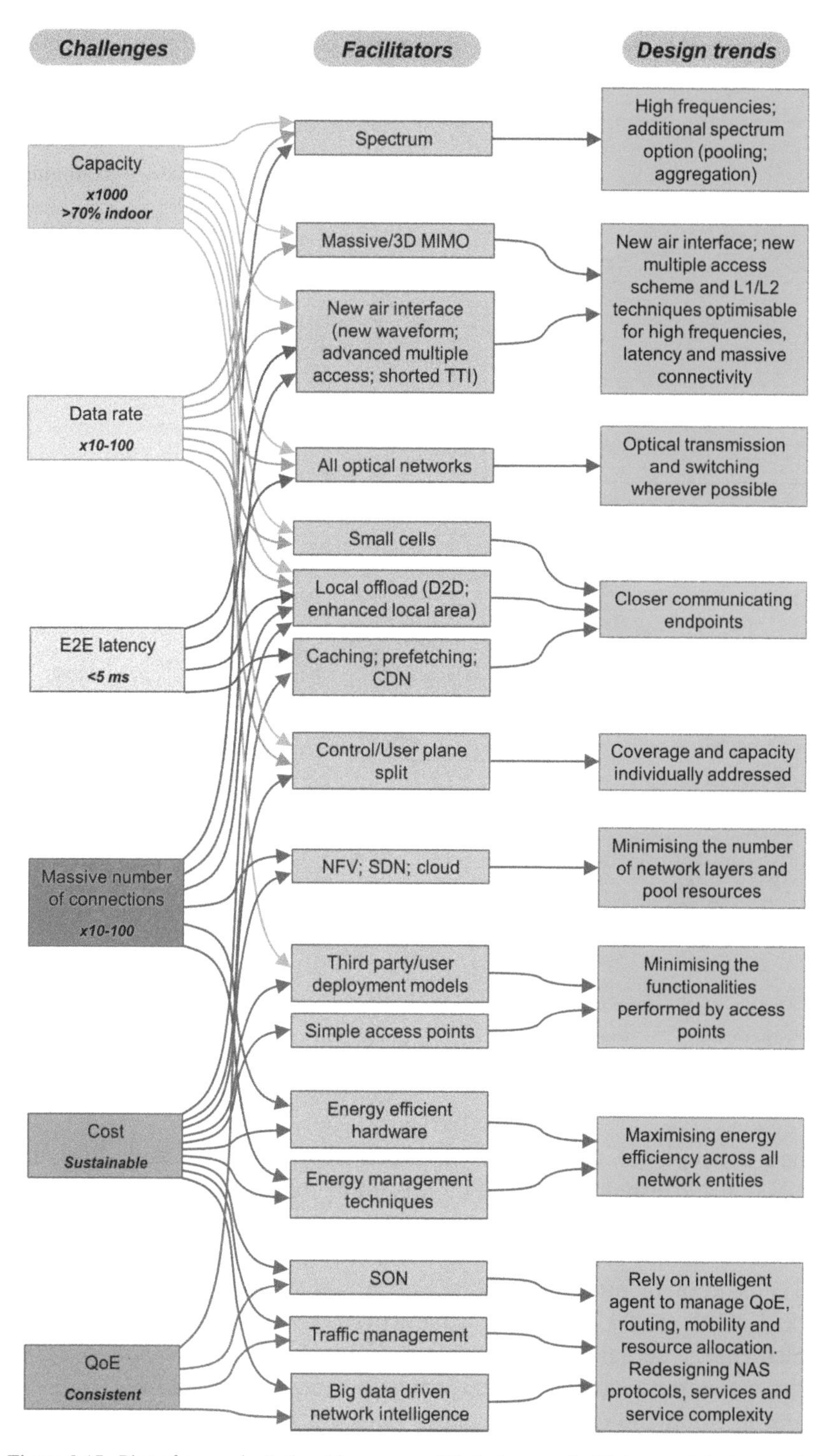

Figure 3.15. Plot of crossed-relationships among 5G challenges, facilitators and design trends.

Table 3.2. List of all the acronyms appearing in figure 3.15.

Acronym	Extended form
CDN	Content Delivery Network
D2D	Device to Device (communication)
E2E	End to End
L1/L2	Layer 1/Layer 2
MIMO	Multiple Input Multiple Output
NAS	Non-Access Stratum
NFV	Network Function Virtualization
QoE	Quality of Experience
SDN	Software Defined Network
SON	Self-Organising Network
TTI	Transmission Time Interval

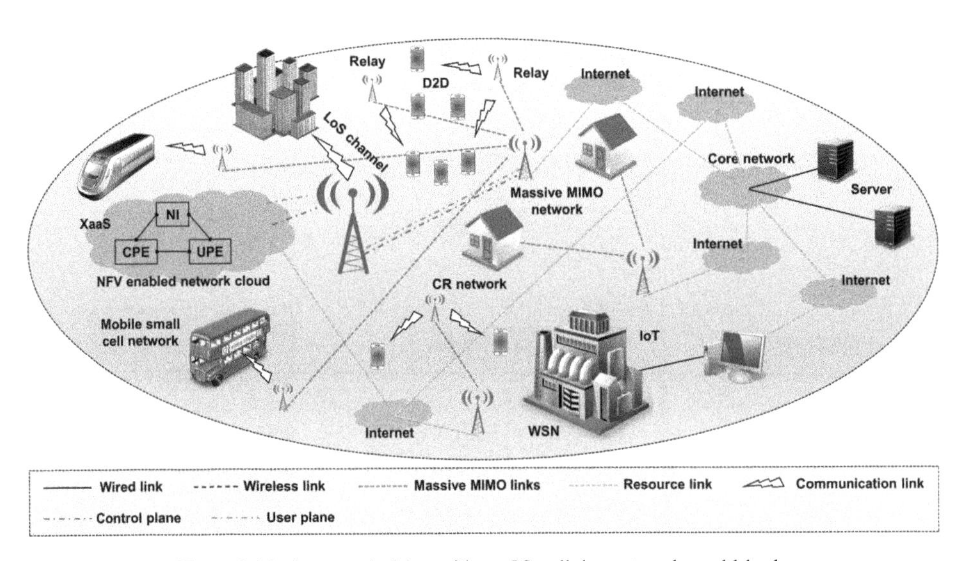

Figure 3.16. A general vision of how 5G cellular network could look.

mentioned concepts of network virtualization, densification and diversification, already being addressed by 4G-LTE, are going to be further developed in the upcoming 5G scenario. In particular, high data rate services are forecasted to be made available to end users by massive MIMO solutions and capillary deployment of small cells, implementing the concepts of Moving Networks (MNs), Ultra-Dense Networks (UDNs) and Ultra-Reliable Networks (URNs). Small cells, in particular, will implement the actual 5G NR paradigm, since they are envisaged to work in the millimetre wave (mm-wave) frequency range, as mentioned below.

Bearing in mind the complex scenario depicted up to now, a few technologies perceived as enabling for 5G are going to be briefly discussed in the following subsections.

Table 3.3. List of all the acronyms appearing in figure 3.16.

Acronym	Extended form
CPE	Control Plane Entity
CR	Cognitive Radio
D2D	Device to Device (communication)
IoT	Internet of Things
LoS	Line of Sight
MIMO	Multiple Input Multiple Output
NFV	Network Function Virtualization
NI	Network Intelligence
UPE	User Plane Entity
WSN	Wireless Sensor Network
XaaS	Network Functionalities as a Service

3.3.1 Massive MIMO

MIMO solutions rely on multiple paths for transmission/reception of signals by means of duplicated antennas arranged in array or matrix fashion. In the current evolution of 4G-LTE, 4×4 MIMO solutions are being investigated and preliminarily deployed. In 5G, massive MIMO solutions will be ventured on, featuring arrays with hundreds of antennas, serving many tens of users at the same time. The reasons why MIMO technology is considered a key enabler of 5G are multiple and acting at diverse levels. Massive MIMO has the capability of improving the radiated energy efficiency up to 100 times and to increase, at the same time, the complexity order capacity of 10 times and more [37]. With the increase of antennas, the energy can be concentrated in small regions and the principle of coherent waveforms superposition can be exploited. Ideally, the wave fronts emitted from the source antenna could add constructively at the intended terminal location and destructively elsewhere, thus reducing interference and energy losses. Moreover, massive MIMO systems can be arranged together with low-power and less costly components. To this regard, MIMO devices require hundreds of inexpensive Power Amplifiers (PAs) in the mW range, rather than ultra-linear very expensive 30-50 W range PAs, currently in use in base stations [37]. In addition, having channels that individually require low-power levels opens the floor to the exploitation of renewable sources and of environmental Energy Harvesting (EH) techniques [38–42]. From a different point of view, massive MIMO can enable latency decrease, through beamforming and fading dips avoidance, as well as simplifying multiple access, due to the combination of OFDM and MIMO [37]. Eventually, MIMO solutions can increase the strength both against unintended interference as well as against intended jamming. If redundant degrees of freedom are available, they can be exploited for cancelling interference or signals generated by intended jammers.

3.3.2 Interference management

As can be easily pictured, with the increase of traffic capacity, network densification and co-use of the same hardware platforms, interference can turn out to be a critical and even impairing aspect for the whole 5G scenario. Therefore, efficient interference management schemes are of vital importance. Two (among other) techniques are briefly mentioned in the following. One solution relies on advance interference management at the receiver [43]. If advanced receivers are available, it could be possible to detect and even decode the symbols belonging to interference within the constellation and the coding scheme of the actual signal to be received. Therefore, the interference signal could be reconstructed and cancelled by the receiver. Another solution to manage interference is the so-called joint scheduling [43]. Basically, additional performance improvements can be enabled by synchronized broadcasting among multiple transmitters scattered in different locations.

3.3.3 Spectrum sharing

Increasing demands driven by 5G in terms of wideband availability will also reflect in spectrum-sharing appropriate techniques in order to be fulfilled. In fact, spectrum sharing can take place according to horizontal or vertical approaches, and 5G will gain from both options. Horizontal spectrum sharing refers to the access of several systems with similar access rights to the same resource, such as, for instance, the Wireless Local Area Network (WLAN). However, vertical spectrum sharing involves several systems with hierarchical access rights [44–45]. Given such a background, there exist two fundamental spectrum sharing techniques of interest for 5G, namely, distributed and centralized solutions [45]. Distributed solutions are meant to exclusively manage transmissions that generate interference and are based on the frequent transmission of generally understood signals across multiple devices, in a listen-before-talk fashion, with the aim of avoiding conflicting activities on the shared spectrum [46]. But centralized spectrum sharing techniques rely on access management operated at a higher hierarchical level, and are more suitable for complex and highly granular networks. An example is the geo-localization database method, in which a database has to be checked in order to obtain information about the resources available at a certain location [47–48]. Another centralized solution is that of the spectrum broker, in which horizontally sharing systems negotiate short term grants to access and use spectrum resources with a central management unit [49–50]. Bearing in mind the spectrum sharing issue from a general perspective, a key enabling technology for its proper solution is Cognitive Radio (CR). It can sense the environment and perform, in dynamic ways, adaption of networking protocols, of spectrum utilization, channel access methods and transmission waveform [44].

3.3.4 Device-to-device communication

D2D communications are categorised at the macro cell level and device level. In the former case, a link is established between a base station and one or more remote devices while, in the latter circumstance, communication takes place directly across devices, i.e. without the involvement/mediation of a base station. 5G will be based upon macro cell level D2D communication protocols, complementing them with device level communication solutions, boosting the overall performance in critical contexts, such as congested areas and at the cell edges [51–52]. Bearing in mind the macro cell level and device level scenarios, four main strategies for D2D communication can be envisaged, taking advantage of different mixing of the just-listed categories, as follows:

1) Device relaying with base station controlled link formation. Let us imagine two mobile terminals laying within the same cell, with one device close to the base station and another at the cell edge. The first device has, of course, a strong link to the radio tower, while the other does not. Both terminals are exchanging data with the base station, but the one at the cell edge is unable to do it efficiently because of the poor link. Given this scenario, D2D communication can take place directly between the two mobile terminals, with the one at the cell edge exploiting the other close to the base station as relay (i.e. bridge) to communicate more efficiently with the radio tower. In this context, both direct D2D and device-to-base station communications are managed and supervised by the radio tower itself (i.e. base station operated control link);
2) Direct D2D communication with base station controlled link formation. In this context, direct D2D communication takes place between two terminals anywhere located within the same cell, with no need to exchange data with the base station. Nonetheless, direct D2D communication is supervised by the base station that manages the control link;
3) Device relaying with device controlled link formation. In this scenario, several mobile terminals, randomly located in space, communicate among themselves. In particular, if one device has to establish a direct link to another terminal, but they are not in sight or the quality of signal is poor, other terminals physically located among them can serve as multiple relays (i.e. bridges) to establish a good connection between the sender and the receiver. In this type of direct D2D communication the control link is also managed by the remote devices themselves. In other words, the base station is not involved in this type of communication;
4) Direct D2D communication with device controlled link formation. Finally, this scenario can be interpreted as a particular (and simplified) case of the context discussed in the previous point. Direct D2D communication takes place between two mobile devices (sender and receiver) that also manage the control link (i.e. base station not involved).

Massive exploitation of direct D2D communications bears the potential to strengthen the quality of service coverage, virtually making it excellent regardless of the location. As a matter of fact, wherever the link to base station is poor, for instance because of physical obstacles, distance to radio tower, or interference, there will be other mobile terminals around that can work as relays. Nevertheless, this poses significant concerns in terms of privacy and security, as user data will travel across devices belonging to other users. In any case, research in the field of M2M communication security will be able to provide effective solutions to mitigate this issue [53–57].

3.3.5 Millimetre wave solutions

The 5G requirements in terms of increased capacity, data rates in the Gbps range, ultra-low latency, and so on, together with the crowded usage of the sub-6 GHz frequency range, are pushing the attention towards the exploitation of solutions in the mm-wave range [58–59]. From the technology point of view, feasibility of mm-wave wireless communications is supported and enabled by significant advances in CMOS [60–62] and MEMS based solutions [63–65]; also concerning high-gain and steerable antennas. On the other hand, the 5G increasing need for pronounced directivity of wireless streams, also to mitigate interference and optimize power consumption as discussed earlier, is pushing in the direction of advanced beamforming capabilities. The exploitation of mm-wave carrier frequencies will enable larger bandwidth allocations, overcoming the presently used 20 MHz channels of 4G-LTE, ensuring at the same time significant reduction of latency. Also, with the increase of frequency, the wavelength reduces, making the scenario of arrays of antennas for massive MIMO with small area occupation possible [66]. As a matter of fact, the mm-wave range also exhibits characteristics making it poorly indicated or unsuitable for cellular communications. Among them, the most critical ones are unquestionably path loss effect [67–69], e.g. due to foliage loss, absorption caused by atmosphere and rain [70–71], poor penetration across obstacles [72–73], objects and barriers like walls. However, mm-wave spectrum is appropriate for short-range transmissions [74], thus enabling all the aforementioned pronounced characteristics that 5G will have to meet. Bearing in mind this scenario, a trend to the way of consolidation is the frequency diversity across the backhaul portion of the network hierarchy. To this regard, a clear frequency divide will characterize 5G networks. The classical macro cells, covering rather extended areas, will mainly work in the sub-6 GHz range. On the other hand, remarkable data throughput, capacity and low latency mentioned above will be enabled via significant network densification. To this end, small cells will be deployed, covering very limited spaces, like a single building or small metropolitan areas (e.g. the lobby of a train station or of a shopping mall). Such small cells will enable massive data transfer working in the mm-wave range, i.e. well above 6 GHz (up to 60–70 GHz), exploiting massive MIMO solutions and advanced beamforming capabilities [75–78].

3.3.6 Other technologies

In order to conclude this section, a few additional technologies regarded as 5G key enablers and not mentioned in the previous subsections, are going to be briefly listed. The items reported below are not to be intended as solutions, bearing limited value with respect to those discussed up to now. On the other hand, they are mentioned here briefly, as an in-depth discussion falls out of the scope of this work. Such technologies can be categorized as follows:

- Ultra-Dense Networks (UDNs). In light of the previously mentioned densification and diversification trends of the 5G network, the characteristics in terms of data rate, accessibility, throughput, and so on, will be relevantly boosted. However, dense and dynamic heterogeneous networks will give rise to challenges related to interference, mobility and backhauling. In order to overcome these issues, the design of new network layer functionalities will be necessary, besides the classical design of physical layers. Such software/algorithm based functions will help smart devices decide how to manage connectivity on the basis of context information [79–80];
- Multi radio access technology association. Still referring to network diversification, 5G will unavoidably lead to integration among different radio access technologies. Apart from sub-6 GHz and mm-wave services, future devices will have to support 3G services, various releases of 4G-LTE, different WiFi standards, as well as D2D communications [81]. To this regard, several ongoing studies aim to address in combined ways the issues of user association and resource allocation [82–83];
- Full duplex radio. An intricate challenge at the technology level is that of enabling full duplex systems, i.e. radios able to transmit and receive simultaneously on the same channel. The availability of such a technology would definitely boost the throughput, but poses at the same time challenging issues in minimizing and possibly removing self-interference between the receiver and the signal generated and transmitted by the same device [84–86];
- Cloud technologies for flexible 5G radio access networks. Eventually, cloud computing and technologies will be increasingly integrated with mobile devices and services, also transcending the classical meaning of cloud. For instance, virtual cloud resources could be aggregated in limited areas by exploiting other mobile devices in the vicinity [87]. Moreover, local cloudlets could be installed in public areas, in order to provide access to mobile terminals in the vicinity with lower latency and wider bandwidth, with respect to classical remote cloud servers [88]. Besides, more and more functionalities that now are executed at the base station level will be virtualized and centrally executed at cloud level, thus boosting flexibility and easing access to the 5G network [89].

Conclusion

In this chapter, comprehensive discussion around mobile communication standards was developed. In the first place, a general background of cellular mobile networks was provided, focusing on the fundamental working principles and on supplementary considerations related to the mobile services ecosystem, listing the most important stakeholders involved (policy makers and private entities). Subsequently, the evolution of mobile standards' generations was discussed, reporting the most relevant innovations introduced by each of them, along with strengths and weaknesses.

The 1st generation (1G), deployed at the turning point of the 1970s and 1980s, brought in the founding innovation of enabling mobile communication among end users. Exclusively voice-based services (i.e. no data) were available and broadcasted as analogue signals. The exploited Frequency Division Multiple Access (FDMA) was limited in terms of spectrum efficiency, mobile terminals were bulky as well as power hungry and broadcasted signals were not encrypted. However, 1G introduced important features like frequency band reuse across cells.

The 2nd generation (2G), deployed from the early 1990s, stepped into the digital world. Digital voice signals could be more efficiently broadcasted by means of the Time Division Multiple Access (TDMA) and non-voice features, as Short Message Service (SMS) started to appear on the landscape. Like 1G, 2G was also not efficient in terms of spectrum utilization and encryption of data was not yet introduced. Nonetheless, a key enabling technology for subsequent generations of mobile services was developed (despite not widely deployed) under the 2G umbrella, namely, the Code Division Multiple Access (CDMA).

The 3rd generation (3G), deployed from 2000, led significant enhancements in terms of performance. First, thanks to CDMA, better spectrum exploitation was possible as well as encryption, thus increasing the security of transmitted data. Then, by means of wider channels and of Evolution-Data Optimized (EV-DO), i.e. channels exclusively dedicated to non-voice services, more and more end users could benefit from capillary voice, messaging and Internet access services.

The 4th generation Long Term Evolution (4G-LTE), deployed from 2010, besides continual increase of performance, started to pursue the concept of flexibility and adaptation of resources and services, based upon end-user demands. To this regard, the possibility of dynamically increasing or decreasing the bandwidth is boosted by Carrier Aggregation (CA) techniques as well as by Orthogonal Frequency Division Multiple Access (OFDMA). 4G-LTE has also started to introduce the multi-path transmission concept, through the use of Multiple Input Multiple Output (MIMO) techniques.

Given the aforementioned scenario, the core topic of the future 5th generation (5G) of mobile communications was then introduced. The main three vectors that 5G will pursue have been mentioned, and they can be listed as follows:

- **EMBB** (Enhanced Mobile Broadband), related to the system capacity growth, with the target of 1000 times increase with respect to current 4G-LTE, leading to 10 Gbps peak and a minimum of 10 Mbps per each user;
- **URLLC** (Ultra Reliable Low Latency Communications), demanded by critical applications, like Vehicle to Vehicle (V2V) communications, aiming at the millisecond latency range;
- **MMTC** (Massive Machine Type Communications), leading to a very large increase of new connections to be supported by 5G services.

Given these main drivers, the most relevant key enabling technologies bearing the potential to make the 5G scenario real, were then listed and discussed in detail. Among them, the most critical seems to be the massive exploitation of MIMO techniques, interference management, spectrum sharing and device-to-device

(D2D) communications. Eventually, apart from the listed items, exploration of 5G New Radio (NR) concepts, through the use of millimetre wave (mm-wave) frequencies and advanced beamforming, appears as a disruptive solution to empower capillary deployment of small cells, bringing impressive network capacity to end users.

References

[1] Tse D and Viswanath P 2005 *Fundamentals of Wireless Communication* 1st edn (Cambridge: Cambridge University Press) p 564
[2] Lee W C Y 2010 *Mobile Communications Design Fundamentals* 2nd edn (Hoboken, NJ: John Wiley) p 398
[3] Bertoni H L 1999 *Radio Propagation for Modern Wireless Systems* 1st edn (Upper Saddle River, NJ: Prentice Hall) p 258
[4] Vidojkovic M 2011 *Configurable Circuits and Their Impact on Multi-Standard RF Front-End Architectures* 1st edn (Eindhoven: Technische Universiteit Eindhoven) p 179
[5] Gupta A and Jha R K 2015 A survey of 5G network: architecture and emerging technologies *IEEE Access* **3** 1206–32
[6] Qualcomm *The Evolution of Mobile Technologies: 1G → 2G → 3G → 4G LTE* https://www.qualcomm.com/documents/evolution-mobile-technologies-1g-2g-3g-4g-lte [accessed 19 June 2017]
[7] Santhi K R, Srivastava V K, SenthilKumaran G and Butare A 2003 Goals of true broad band's wireless next wave (4G-5G) *Proc. IEEE Vehicular Tech. Conf. VTC (Orlando, FL, Oct. 2003)* pp 2317–21
[8] Halonen T, Romero J and Melero J (ed) 2003 *GSM, GPRS and EDGE Performance: Evolution Towards 3G/UMTS* 2nd edn (NewYork: Wiley) p 654
[9] Gammel P, Pehlke D R, Brunel D, Kovacic S J and Walsh K *5G in Perspective: A Pragmatic Guide to What's Next* http://www.skyworksinc.com/Products_5G_Whitepaper.aspx?source=home [accessed 20 June 2017]
[10] Telxpert *A Roadmap For Cellular Networks Evolution Towards LTE-Advance Networks* http://www.telxperts.com/lte/roadmap-cellular-networks-evolution-towards-lte-advance-networks/ [accessed 20 June 2017]
[11] Sesia S, Toufik I and Baker M 2009 *LTE—The UMTS Long Term Evolution: From Theory to Practice* 1st edn (Hoboken, NJ: John Wiley) p 648
[12] Qualcomm LTE *TDD: The Global Solution For Unpaired Spectrum* https://www.qualcomm.com/documents/lte-tdd-global-solution-unpaired-spectrum [accessed 19 June 2017]
[13] Ref Ai B, Guan K, He R, Li J, Li G, He D, Zhong Z and Huq K M S 2017 On indoor millimeter wave massive MIMO channels: measurement and simulation *IEEE J. Sel. Areas Commun.* **35** 1678–90
[14] Sipal D, Abegaonkar M P and Koul S K 2017 Easily extendable compact planar UWB MIMO antenna array *IEEE Antennas Wirel. Propag. Lett.* **16** 2328–31
[15] Nguyen V G, Do T X and Kim Y 2016 SDN and virtualization-based LTE mobile network architectures: a comprehensive survey *Wirel. Pers. Commun.* **86** 1401–38
[16] Zaki Y, Zhao L, Goerg C and Timm-Giel A 2011 LTE mobile network virtualization *J. Mobile Netw. Appl.* **16** 424–32

[17] Bhushan N, Li J, Malladi D, Gilmore R, Brenner D R, Damnjanovic A D, Sukhavasi R T, Patel C and Geirhofer S 2014 Network densification: the dominant theme for wireless evolution into 5G *IEEE Commun. Mag.* **52** 82–9
[18] Andrews J G, Zhang X, Durgin G D and Gupta A K 2016 Are we approaching the fundamental limits of wireless network densification? *IEEE Commun. Mag.* **54** 184–90
[19] Nguyen V M and Kountouris M 2017 Performance limits of network densification *IEEE J. Sel. Areas Commun.* **35** 1294–308
[20] Zhao C, Huang L, Zhao Y and Du X 2017 Secure machine-type communications toward LTE heterogeneous networks *IEEE Wirel. Commun.* **24** 82–7
[21] Slamnik N, Okic A and Musovic J 2016 Conceptual radio resource management approach in LTE heterogeneous networks using small cells number variation *Proc. Int. Symp. on Telecommun. BIHTEL (Sarajevo, Oct. 2016)* pp 1–5
[22] Petrut I, Otesteanu M, Balint C and Budura G 2016 On the uplink performance in LTE heterogeneous network *Proc. Int. Conf. on Commun. COMM (Bucharest, June 2016)* pp 191–4
[23] Elfadil H E, Ali M A I and Abas M 2015 Performance evaluation of heterogeneous networks schemes in LTE networks *Proc. Int. Conf. on Computing, Control, Networking, Electron. and Embedded Syst. Eng. ICCNEEE (Khartoum, Sept. 2015)* pp 401–8
[24] Rysavy Research *LTE and 5G Innovation: Igniting Mobile Broadband* http://www.rysavy.com/Articles/2015-08-Rysavy-4G-Americas-LTE-5G-Innovation.pdf [accessed 22 June 2017]
[25] Econocom *How the Internet of Things is Revolutionising Business Models* https://blog.econocom.com/en/blog/how-the-internet-of-things-is-revolutionising-business-models/ [accessed 23 June 2017]
[26] Vermesan O and Friess P (ed) 2014 *Internet of Things Applications—From Research and Innovation to Market Deployment* 1st edn (Aalborg: River Publishers) p 364
[27] Uckelmann D, Harrison M and Michahelles F (ed) 2011 *Architecting the Internet of Things* 1st edn (Berlin: Springer) p 353
[28] MachNation *The Development of the Internet of Everything* https://d175jir5ufcz29.cloudfront.net/legacy-mn-wp-media/2014/06/MachNation-Development-of-the-IoE.pdf [accessed 23 June 2017]
[29] ITU *The Tactile Internet* https://www.itu.int/dms_pub/itu-t/oth/23/01/T23010000230001PDFE.pdf [accessed 23 June 2017]
[30] Xiang W, Zheng K and Shen X S (ed) 2017 *5G Mobile Communications* 1st edn (Cham: Springer) p 691
[31] Ericsson *Evolving LTE to Fit the 5G Future* https://www.ericsson.com/en/publications/ericsson-technology-review/archive/2017/evolving-lte-to-fit-the-5g-future [accessed 23 June 2017]
[32] Huawei *5G Network Architecture a High-Level Perspective* http://www.huawei.com/minisite/hwmbbf16/insights/5G-Nework-Architecture-Whitepaper-en.pdf [accessed 23 June 2017]
[33] 3GPP *The Path to 5G: as much Evolution as Revolution* http://www.3gpp.org/news-events/3gpp-news/1774-5g_wiseharbour [accessed 23 June 2017]
[34] Agyapong P K, Iwamura M, Staehle D, Kiess W and Benjebbour A 2014 Design considerations for a 5G network architecture *IEEE Commun. Mag.* **52** 65–75
[35] PPP in Horizon 2020 *Advanced 5G Network Infrastructure for the Future Internet*—Creating a Smart Ubiquitous Network for the Future Internet https://5g-ppp.eu/wp-content/uploads/2014/02/Advanced-5G-Network-Infrastructure-PPP-in-H2020_Final_November-2013.pdf [accessed 28 June 2017]

[36] Osseiran A, Boccardi F, Braun V, Kusume K, Marsch P, Maternia M, Queseth O, Schellmann M, Schotten H, Taoka H, Tullberg H, Uusitalo M A, Timus B and Fallgren M 2014 Scenarios for 5G mobile and wireless communications: the vision of the METIS project *IEEE Commun. Mag.* **52** 26–35

[37] Larsson E G, Edfors O, Tufvesson F and Marzetta T L 2014 Massive MIMO for next generation wireless systems *IEEE Commun. Mag.* **52** 186–95

[38] Ulukus S, Yener A, Erkip E, Simeone O, Zorzi M, Grover P and Huang K 2015 Energy harvesting wireless communications: a review of recent advances *IEEE J. Sel. Areas Commun.* **33** 360–81

[39] Iannacci J, Sordo G, Serra E and Schmid U 2016 The MEMS four-leaf clover wideband vibration energy harvesting device: design concept and experimental verification *Microsyst. Technol.* **22** 1865–81

[40] Iannacci J and Sordo G 2016 Up-scaled macro-device implementation of a MEMS wideband vibration piezoelectric energy harvester design concept *Microsyst. Technol.* **22** 1639–51

[41] Iannacci J, Serra E, Di Criscienzo R, Sordo G, Gottardi M, Borrielli A, Bonaldi M, Kuenzig T, Schrag G, Pandraud G and Sarro P M 2014 Multi-modal vibration based MEMS energy harvesters for ultra-low power wireless functional nodes *Microsyst. Technol.* **20** 627–40

[42] Iannacci J 2017 Microsystem based energy harvesting (EH-MEMS): powering pervasivity of the Internet of Things (IoT)—A review with focus on mechanical vibrations *J. King Saud Univ.* 1–9

[43] Nam W, Bai D, Lee J and Kang I 2014 Advanced interference management for 5G cellular networks *IEEE Commun. Mag.* **52** 52–60

[44] Wyglinski A M, Nekovee M and Hou T (ed) 2009 *Cognitive Radio Communications and Networks: Principles and Practice* 1st edn (Amsterdam: Academic Press) p 736

[45] Irnich T, Kronander J, Selén Y and Li G 2013 Spectrum sharing scenarios and resulting technical requirements for 5G systems *Proc. IEEE Int. Symp. on Personal, Indoor and Mobile Radio Commun. PIMRC (London, Sept. 2013)* pp 127–32

[46] Buchwald G J, Kuffner S L, Ecklund L M, Brown M and Callaway E H Jr. 2008 The design and operation of the IEEE 802.22.1 disabling beacon for the protection of TV whitespace incumbents *Proc. IEEE Symp. on New Frontiers in Dynamic Spectrum Access Networks (Chicago, IL, Oct. 2008)* pp 1–6

[47] Electronic Communications Committee (ECC) *Technical and Operational Requirements for the Possible Operation of Cognitive Radio Systems in the 'White Spaces' of the Frequency Band 470–790 MHz* http://www.erodocdb.dk/docs/doc98/official/pdf/ECCRep159.pdf [accessed 5 July 2017]

[48] Federal Communications Commission (FCC) *Third Memorandum Opinion and Order in the Matter of Unlicensed Operation in the TV Broadcast Bands (ET Docket No. 04-186) and Additional Spectrum for Unlicensed Devices Below 900 MHz and in the 3 GHz Band* (ET Docket No. 02-380) https://apps.fcc.gov/edocs_public/attachmatch/FCC-12-36A1_Rcd.pdf [accessed 5 July 2017]

[49] Huang J, Berry R A and Honig M L 2006 Auction-based spectrum sharing *Mob. Netw. Appl.* **11** 405–18

[50] Shen F, Li D, Lin P H and Jorswieck E 2015 Auction based spectrum sharing for hybrid access in macro-femtocell networks under QoS requirements *Proc. IEEE Int. Conf. on Commun. ICC (London, June 2015)* pp 3335–40

[51] Tehrani M N, Uysal M and Yanikomeroglu H 2014 Device-to-device communication in 5G cellular networks: challenges, solutions, and future directions *IEEE Commun. Mag.* **52** 86–92
[52] Orsino A, Gapeyenko M, Militano L, Moltchanov D, Andreev S, Koucheryavy Y and Araniti G 2015 Assisted handover based on device-to-device communications in 3GPP LTE systems *Proc. IEEE Globecom Workshops (San Diego, CA, Dec. 2015)* pp 1–6
[53] Cha I, Shah Y, Schmidt A U, Leicher A and Meyerstein M V 2009 Trust in M2M communication *IEEE Veh. Tech. Mag.* **4** 69–75
[54] Yue J, Ma C, Yu H and Zhou W 2013 Secrecy-based access control for device-to-device communication underlaying cellular networks *IEEE Commun. Lett.* **17** 2068–71
[55] Perrig A, Stankovic J and Wagner D 2004 Security in wireless sensor networks *ACM J. Commun.* **47** 53–7
[56] Zhou Y, Fang Y and Zhang Y 2008 Securing wireless sensor networks: a survey *IEEE Commun. Surv. Tutor* **10** 6–28
[57] Muraleedharan R and Osadciw L 2006 Jamming attack detection and countermeasures in wireless sensor network using ant system *Proc. SPIE* **6248** 1–12
[58] Rappaport T S, Sun S, Mayzus R, Zhao H, Azar Y, Wang K, Wong G N, Schulz J K, Samimi M and Gutierrez F 2013 Millimeter wave mobile communications for 5G cellular: it will work! *IEEE Access* **1** 335–49
[59] Pi Z and Khan F 2011 An introduction to millimeter-wave mobile broadband systems *IEEE Commun. Mag.* **49** 101–7
[60] Gutierrez F, Agarwal S, Parrish K and Rappaport T S 2009 On-chip integrated antenna structures in CMOS for 60 GHz WPAN systems *IEEE J. Sel. Areas Commun.* **27** 1367–78
[61] Rappaport T S, Ben-Dor E, Murdock J N and Qiao Y 2012 38 GHz and 60 GHz angle-dependent propagation for cellular & peer-to-peer wireless communications *Proc. IEEE Int. Conf. on Commun. ICC (Ottawa, June 2012)* pp 4568–73
[62] Rappaport T S, Murdock J N and Gutierrez F 2011 State of the art in 60-GHz integrated circuits and systems for wireless communications *Proc. IEEE* **99** 1390–436
[63] Iannacci J, Tschoban C, Reyes J, Maaß U, Huhn M, Ndip I and Pötter H 2016 RF-MEMS for 5G mobile communications: A basic attenuator module demonstrated up to 50 GHz *Proc. IEEE SENSORS (Orlando, FL, Oct.–Nov. 2016)* pp 1–3
[64] Iannacci J, Huhn M, Tschoban C and Pötter H 2016 RF-MEMS technology for future mobile and high-frequency applications: reconfigurable 8-bit power attenuator tested up to 110 GHz *IEEE Electron Device Lett.* **37** 1646–9
[65] Iannacci J, Huhn M, Tschoban C and Pötter H 2016 RF-MEMS technology for 5G: series and shunt attenuator modules demonstrated up to 110 GHz *IEEE Electron Device Lett.* **37** 1336–9
[66] Rusek F, Persson D, Lau B K, Larsson E G, Marzetta T L, Edfors O and Tufvesson F 2013 Scaling up MIMO: opportunities and challenges with very large arrays *IEEE Signal Process. Mag.* **30** 40–60
[67] Al-Samman A M, Rahman T A, Azmi M H, Zulkefly N R and Mataria A M S 2016 Path loss model for outdoor environment at 17 GHz mm-wave band *Proc. IEEE Int. Colloquium on Signal Processing & its Applications CSPA (Malacca City, March 2016)* pp 179–82
[68] Sarma A D, Prasad M V S N and Pandit S N N 1999 Modelling of path loss for land mobile cm and mm wave communication systems *Proc. Int. Conf. on Electromagnetic Interference and Compatibility (New Delhi, Dec. 1999)* pp 75–8

[69] Ravindra K and Sarma A D 1999 Modelling path loss in the near field region for cm and mm wave mobile communication *Proc. Int. Conf. on Electromagnetic Interference and Compatibility (New Delhi, Dec. 1999)* pp 79–82

[70] Serov E A, Koshelev M A, Parshin V V and Tretyakov M Y 2010 Atmosphere continuum absorption investigation at MM waves *Proc. Int. Kharkov Symp. on Phys. and Eng. of Microwaves, Millimeter and Submillimeter Waves (Kharkiv, June 2010)* pp 1–3

[71] Lehto A, Tuovinen J and Raisanen A 1991 Proposed method for measurement of absorption loss and beam efficiency of large MM-wave reflector antennas in compact antenna test range by using hot and cold loads *IEE Proc. H - Microwaves, Antennas and Propagation* **138** 13–8

[72] Capone A, Filippini I, Sciancalepore V and Tremolada D 2015 Obstacle avoidance cell discovery using mm-waves directive antennas in 5G networks *Proc. IEEE Annual Int. Symp. on Personal, Indoor, and Mobile Radio Commun. PIMRC (Hong Kong, Aug.–Sept. 2015)* pp 2349–53

[73] Simic L, Panda S, Riihijarvi J and Mahonen P 2017 Coverage and robustness of mm-wave urban cellular networks: multi-frequency HetNets are the 5G future *Proc. Ann. IEEE Int. Conf. on Sensing, Commun., and Networking SECON (San Diego, CA, June 2017)* pp 1–9

[74] International Telecommunication Union (ITU) *Estimated Spectrum Bandwidth Requirements for the Future Development of IMT-2000 and IMT-Advanced* http://www.itu.int/pub/R-REP-M.2078 [accessed 6 July 2017]

[75] Krasko O, Brych M, Masyuk A and Klymash M 2016 Flexible backhaul architecture for densely deployed 5G small cells based on OWTDMA network *Proc. Third Int. Scientific-Practical Conf. Problems of Infocommunications Science and Technology PIC S&T (Kharkiv, Oct. 2016)* pp 33–5

[76] Hou X, Wang X, Jiang H and Kayama H 2016 Investigation of massive MIMO in dense small cell deployment for 5G *Proc. IEEE Vehicular Technol. Conf. VTC-Fall (Montreal, Sept. 2016)*

[77] Behnad A and Wang X 2017 Virtual small cells formation in 5G networks *IEEE Commun. Lett.* **21** 616–9

[78] Nasr A I and Fahmy Y 2016 Millimeter-wave wireless backhauling for 5G small cells: Star versus mesh topologies *Proc. Int. Conf. on Microelectron. ICM (Cairo, Dec. 2016)* pp 85–8

[79] Yu G, Xu L, Feng D, Yin R, Li G Y and Jiang Y 2014 Joint mode selection and resource allocation for device-to-device communications *IEEE Trans. Commun.* **62** 3814–24

[80] Wang L and Wu H 2014 Fast pairing of device-to-device link underlay for spectrum sharing with cellular users *IEEE Commun. Lett.* **18** 1803–6

[81] Andrews J G, Buzzi S, Choi W, Hanly S V, Lozano A, Soong A C K and Zhang J C 2014 What will 5G be *IEEE J. Sel. Areas Commun.* **32** 1065–82

[82] Ye Q, Rong B, Chen Y, Al-Shalash M, Caramanis C and Andrews J G 2013 User association for load balancing in heterogeneous cellular networks *IEEE Trans. Wirel. Commun.* **12** 2706–16

[83] Corroy S, Falconetti L and Mathar R 2012 Cell association in small heterogeneous networks: Downlink sum rate and min rate maximization *Proc. IEEE Wirel. Commun. and Networking Conf. WCNC (Paris, April 2012)* pp 888–92

[84] Kim K, Jeon S W and Kim D K 2015 The feasibility of interference alignment for full-duplex MIMO cellular networks *IEEE Commun. Lett.* **19** 1500–3

[85] Wang K, Zhang R, Zhong Z, Zhang X and Pang X 2017 Measurement of self-interference channels for full-duplex relay in an urban scenario *Proc. IEEE Int. Conf. on Commun. ICC (Paris, May 2017)* pp 1153–8

[86] Shojaeifard A, Wong K K, Di Renzo M, Zheng G, Hamdi K A and Tang J 2017 Full-duplex versus half-duplex large scale antenna system *Proc. IEEE Int. Conf. on Commun. ICC (Paris, May 2017)* pp 743–8

[87] Marinelli E E *Hyrax: Cloud Computing on Mobile Devices using MapReduce* http://reports-archive.adm.cs.cmu.edu/anon/2009/CMU-CS-09-164.pdf [accessed 6 July 2017]

[88] Satyanarayanan M, Bahl P, Caceres R and Davies N 2009 The case for VM-based cloudlets in mobile computing *IEEE Pervasive Comput.* **8** 14–23

[89] Rost P, Bernardos C J, De Domenico A, Di Girolamo M, Lalam M, Sabella D and Wübben D 2014 Cloud technologies for flexible 5G radio access networks *IEEE Commun. Mag.* **52** 68–76

IOP Publishing

RF-MEMS Technology for High-Performance Passives
The challenge of 5G mobile applications
Jacopo Iannacci

Chapter 4

RF-MEMS passives for 5G applications: Case studies and design examples

4.1 An introduction on RF passives' specifications trends in 5G

The insight provided in previous chapters on RF-MEMS technologies is quite exhaustive. On one hand, an achievable performance along with the added value of pronounced reconfigurability/tunability, targetable through RF passive components designed in MEMS technology, were reported and discussed in the first chapter. On the other hand, trends in terms of system-level performance, triggered by current 4G-LTE mobile communications and destined to be pursued relentlessly with the advent of 5G standards, have been discussed in a generous amount of detail in the preceding chapter. To complete the picture, an intermediate chapter, between the RF-MEMS introductory and that of 5G, unfolded the main reasons that, in the past, limited concrete market absorption of RF passive products based on MEMS technology, and framed also the significantly changed scenario in which RF-MEMS are on the way to consolidation, especially with reference to 4G-LTE market applications. In light of this landscape, the purpose of this last chapter is to blend together all the elements scattered around, up to here, and build up more technical insight around the possible exploitation of RF-MEMS technology in the 5G frame of reference.

As a matter of fact, performance indicators and trends discussed throughout the chapter on 5G were mainly lying on the system-level plane of abstraction. In addition, the key enabling technologies listed before, like massive Multiple Input Multiple Output (MIMO), millimetre wave (mm-wave) solutions, and so on, refer to complete systems and sub-systems. Therefore, prior to stepping further into the area of RF-MEMS design and simulation technicalities, effort should be oriented towards lowering specifications and characteristics from the system-level down to the component-level.

Bearing in mind the aforementioned target, a first useful step consists in listing classes of RF passives that are expected to be beneficial in order to empower the 5G

doi:10.1088/978-0-7503-1545-6ch4

scenario discussed before. Such components, also taking into account the 5G New Radio (NR) concepts that will lie in the mm-wave range, can be grouped as follows:

1) Very-wideband switches and switching units (e.g. Multiple Pole Multiple Throw—MPMT) with low-loss (when ON), high-isolation (when OFF) and very-low adjacent channels cross-talk, working from 2–3 GHz up to 60–70 GHz (and more);
2) Reconfigurable filters with pronounced stopband rejection characteristics and very-low attenuation of the passed band;
3) Very-wideband multi-state impedance tuners;
4) Programmable step attenuators with multiple configurations and very flat characteristics over 60–70 GHz frequency spans;
5) Very-wideband multi-state/analogue phase shifters;
6) Hybrid devices with mixed phase shifting and programmable attenuation—functionalities described in points 4) and 5) blended into a unique device;
7) Miniaturized antennas and arrays of antennas, possibly integrated monolithically with one or more of the devices described in the previous points from 1) to 6).

These listed devices could find application in several parts of the 5G intricate landscape, both with reference to smartphones and end user terminals, as well as to fixed/mobile infrastructure. For instance, high-performance switches and switching units would be beneficial to MIMO systems, both on the terminal and infrastructure ends. In addition, wideband impedance tuners are critical devices to enable adaptive matching between the antenna and RF Front End (RFFE) of smartphones. From a different perspective, mm-wave small cells will require advanced beamforming capability. To this end, widely reconfigurable phase shifters, as well as multi-state programmable step attenuators, are critical components to ensure proper feeding of the array of antennas. Moreover, it would be desirable to implement both of these functions within the same component (monolithic solution), thus reducing losses, area occupation, hardware complexity and cost. Following the same direction, monolithic realization of the array of antennas and of the passive front end necessary for advanced beamforming, i.e. phase shifters and attenuators, would mean stepping to a higher level from the point of view of system performance.

Given the just disclosed background on RF passives of interest for 5G applications, it is now time to move our attention to the expected performance figures and characteristics. To this regard, the *performance bowl* is shown in figure 4.1.

Within the bowl, each *bubble* is a performance indicator and its placement along the vertical axis indicates if it is expected to be high/wide (bowl ceiling), or low/small (bowl floor). For instance, isolation of switching elements (when OFF) is supposed to be very pronounced, while cross-talks between adjacent channels must be kept as low as possible. These are the reasons why the isolation and cross-talk bubbles are to the top and bottom of the bowl height, respectively.

Given the specifications within the performance bowl in figure 4.1, it is not viable to quantify the unique numbers to be targeted, as they strongly depend on multiple factors, such as the frequency range of operation, as well as the requirements

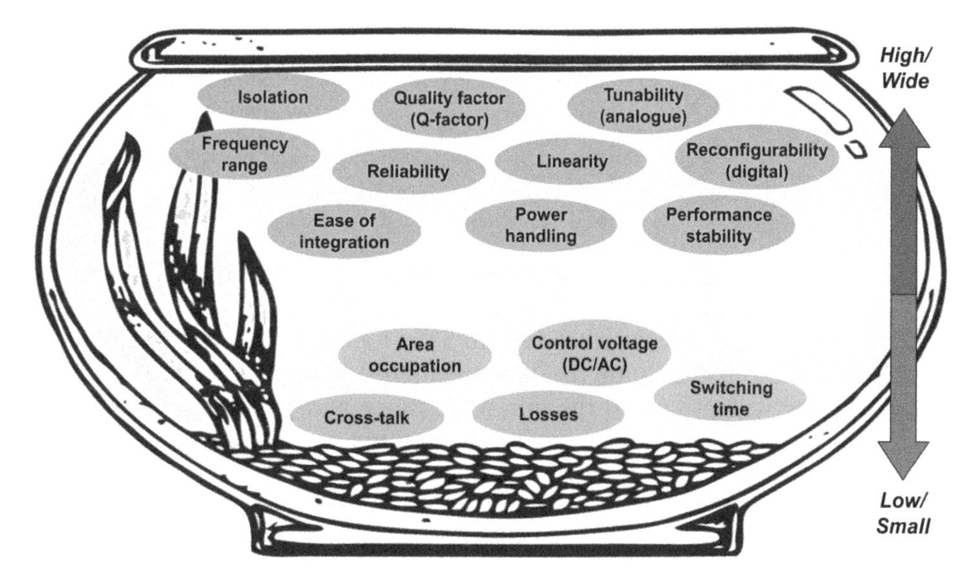

Figure 4.1. *Performance bowl* including the most critical specifications and characteristics expected from RF passives for 5G applications. The vertical placement of each *bubble* (characteristic) within the bowl depends on how high or low it is expected to be (e.g. high-isolation; low-losses).

imposed by the specific application. In addition, the topmost targetable performance for a certain device also depends on its complexity. For example, if the losses of a single switch (when ON) can reasonably approach a certain number, the same characteristic of a switching unit with multiple cascaded stages will be unavoidably worse than that value. In light of these considerations, it is in any case possible to sketch a quantitative list of reference target numbers for some of the performance indicators reported in figure 4.1, bearing in mind the scenario of 5G.

- Frequency range. In light of the discussion developed before, in the 5G scenario there will be significant diversification of the network infrastructure. To this regard, there clearly emerged the frequency divide between the classical backhaul infrastructure working in the sub-6 GHz range, compared against small cells operating in the mm-wave range. Therefore, RF passives will be required to exhibit good performance in a very-wide span, starting from 2–3 GHz and ramping up to 60–70 GHz. This is particularly critical for components to be integrated within smartphones and mobile terminals in general, as such devices will have to work properly when linked both to traditional base stations (sub-6 GHz) and to small cells (mm-wave range);
- Isolation. Given the pronounced RFFEs hardware reconfigurability and switching capabilities that need to be addressed within the 5G scenario, a critical feature of switches and, more in general, switch-based passives, will be isolation (when OFF), which is expected to be as high as possible. Just to provide an indication, isolation better than −30/−40 dB for frequencies as high as possible, would be desirable. Of course, if it is a reasonable task to design micro-relays with such an isolation operating up to a few GHz, the

target becomes much more challenging when dealing with mm-wave, because of input/output parasitic coupling effects;

- Losses. Referring to the scenario of the previous point, the other side of the coin is that of losses introduced by switches (when ON) and, more in general, RF passives when operating in non-isolating configurations. It is straightforward that losses are to be as limited as possible. It would be desirable to keep losses below −1 dB on the widest possible frequency range. As a matter of fact, however, losses get worse when the frequency of operation increases, and also when the passive is longer/more complex, because of parasitic effects. The challenge of keeping the losses to be as limited as possible is particularly arduous; as in 5G both the aforementioned factors will take place, i.e. frequency will rise up to the mm-wave range and passives will have to be widely reconfigurable, that equals having complex, articulated and redundant designs;
- Cross-talk. Still bearing in mind the pronounced reconfigurability of 5G, the cross-talk emerges as another critical performance figure. It can be defined as the extent to which adjacent channels, ideally conceived to be completely isolated from each other, in fact interfere and couple one another. It is easy to embrace how critical it is to keep the cross-talk low in multi-path scenarios, like the one of massive MIMO technologies and advanced beamforming capability. To this regard, a desirable threshold would keep cross-talks below −50/−60 dB over the widest frequency range possible; although it tends to degrade ramping up to the mm-wave range due to the kicking in of parasitic effects and unwanted couplings;
- Switching time. Moving our attention towards performance indicators not directly related to the RF and electromagnetic characteristics of passives, it is worth mentioning the switching time. Given the pronounced reconfigurability demanded of 5G RFFEs and RF hardware, it clearly emerges that idle time in commuting from one state to another must be kept as limited as possible. Just to provide an example, this could be the case of a mobile device hopping from sub-6 GHz coverage to mm-wave small cell services, during an outdoor to indoor environment transition. Breaking down the commutation time issue from system to component level, the fundamental critical device is the single switch and the required time for its transition from stable OFF to stable ON configuration, as well as vice-versa. Given the millisecond-range latency targeted by 5G services, the basic relay switching time has to be definitely lower than 1 ms, with few fractions of ms (e.g. 200–300 μs) as a reasonable target;
- Control voltage. Eventually, still referring to switches and RF passive reconfigurable devices, their operation requires driving them by means of suitable DC/AC electrical signals, i.e. voltages/currents. It is clear that, if within the reference scenario of ground infrastructures, providing biasing signals is, in principle, not a critical issue, the context changes dramatically in mobile and battery-operated devices. In the latter case, the extent of DC/AC biasing signals, as well as power consumption in driving RF passives, should

be kept as low as possible. It would also be desirable to maintain compliance with CMOS typical driving voltages, in order to avoid incorporation of additional ad hoc voltage pumping circuitry. This means working with control voltages not higher than 1–2 V.

In the following sections, multiphysics electromechanical and electromagnetic modelling of RF-MEMS design case studies is going to be discussed, targeting the just-reported 5G driven performance indicators and characteristics.

4.2 A few notes about the RF-MEMS technology platform

Prior to discussing the technical details of a few case studies and reference designs of RF-MEMS for 5G applications, some references are going to be dispensed concerning the technology fabrication process. In order to keep the understanding of the reported examples simple and coherent, a unique RF-MEMS manufacturing process is chosen, it being the 6-inch silicon wafers RF-MEMS surface micromachining process flow, available at the Center for Materials and Microsystems (CMM) of Fondazione Bruno Kessler (FBK) in Trento (Italy). The main fabrication steps of the CMM-FBK RF-MEMS technology are schematically depicted by the simplified sequence in figure 4.2.

The technology features two buried conductive layers, namely polycrystalline silicon and aluminium, meant for DC biasing/calibrated resistive loads and RF signals underpasses, respectively. Subsequently, a thin gold layer is evaporated above the opened areas to the underneath aluminium layer, in order to improve the quality of the ohmic contact. Then, a photoresist sacrificial layer is patterned wherever MEMS elevated membranes are supposed to be. Finally, two distinct gold electrodeposition steps define the structural parts of the device, and air-gaps are eventually released by removing the sacrificial layer. The most relevant geometrical, mechanical and electrical properties of the layers are reported in table 4.1, while a more detailed discussion around the CMM-FBK RF-MEMS technology is available in [1–3].

4.3 Electromechanical simulation of RF-MEMS devices

The target of this section is to describe a fundamental methodology devoted to the modelling and simulation of the coupled electromechanical behaviour of RF-MEMS devices. Before discussing practical cases, however, some considerations must be developed. First, the approach that is going to be shown is neither the only available, nor the best one could think to exploit in any situation, when developing a certain RF-MEMS project. Such a circumstance takes place because it is the specific problem to be addressed and solved the fundamental driver defining which simulation approach can be indicated to be more suitable than the others available. For the purposes of clear understanding, let us just mention a reference example. If the RF-MEMS designer/developer/engineer has to develop a more structured comprehension of the materials behaviour against the amount of RF power the device has to stand, or of the metal-to-metal contact degradation versus cycling,

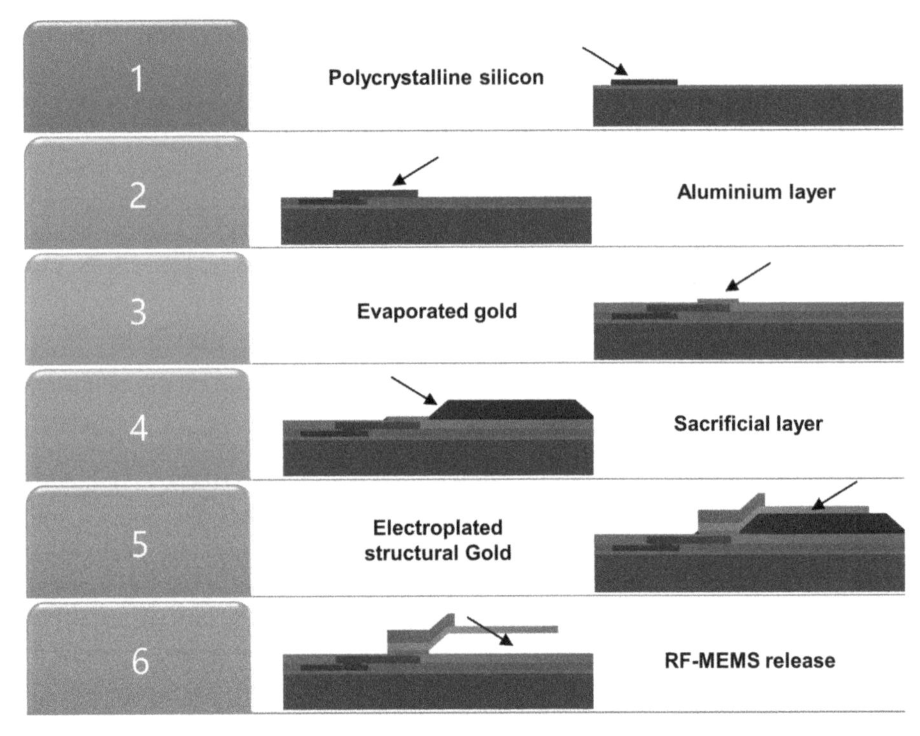

Figure 4.2. Schematic cross-sectional simplified sequence of the main fabrication steps of the CMM-FBK RF-MEMS surface micromachining process. Details about the deposited layers are reported in table 4.1.

Table 4.1. Mechanical and electrical fundamental properties of the layers composing the CMM-FBK RF-MEMS surface micromachining fabrication flow depicted in figure 4.2.

Physical layer	Fabrication step (see figure 4.2)	Nominal thickness	Sheet resistance [Ω/sq]	Dielectric constant [F/m]	Young's modulus [GPa]
Silicon substrate		625 μm	>5000 Ω. cm	11.9	
Field oxide		1 μm		3.94	
Polycrystalline silicon	1	630 nm	1584		
Insulating silicon oxide		300 nm		3.94	
Aluminium layer	2	630 nm	0.0654		
Insulating silicon oxide		100 nm		3.94	
Evaporated gold	3	150 nm	0.126		
Sacrificial layer	4	3 μm			
Electroplated structural gold (layer 1)	5	2 μm	0.022		98.5
Electroplated structural gold (layer 2)	5	3 μm	0.0055		98.5

simulation tools dedicated to the mechanical/structural analysis of materials, possibly based on the Finite Element Method (FEM) approach, are more indicated than others. On the other hand, if the designer/developer/engineer needs to evaluate device-level characteristics of a certain RF-MEMS component placed within a sub-system also featuring active electronics, it is definitely more appropriate to rely on fast simulation tools. The latter are able to describe the overall multiphysics behaviour of the RF passive in MEMS technology, also ensuring compatibility and proper interface to the CMOS part, so that hybrid (CMOS/RF-MEMS) devices can be simulated as a whole.

Whereas, on one hand, FEM tools address very accurate results and predictions, on the other hand they require more computational capacity, and are not suitable to describe the overall electromechanical, RF and electronic coupled behaviour within the same framework. In contrast, fast tools based on analytical and/or simplified (order reduced) models do not target the same accuracy and flexibility of FEM simulation environments, although they do enable easy identification and evaluation of multiple trade-offs existing in RF-MEMS. Moreover, as mentioned above, they are compatible with standard CMOS design-kits and, therefore, can be simulated within the same analysis frameworks suited for electronics development.

Of course, using more accurate or high-level modelling tools is not an exclusive choice, but rather a matter of selecting the proper approach depending on the specific needs arising at different stages of the RF-MEMS design concept development chain. For the purposes of this work, case studies will be reported at device level, in order to emphasise the correlation between changes/choices at layout geometry level and the performance indicators mentioned at the beginning of this chapter. With the aim of maintaining easy understanding of the reported examples to the widest possible extent, specific technical details regarding simulations/simulators settings will be omitted. Nonetheless, references to articles and books dealing with such technicalities will be provided, thus facilitating readers who are interested in carrying out more detailed studies on specific topics of RF-MEMS devices design and development.

4.3.1 Static structural simulation of the pull-in/pull-out characteristic

Bearing in mind the list of critical performances requested to RF passive from high-level 5G specifications outlined in the beginning of this chapter, our attention is now going to be focused on the pull-in voltage of RF-MEMS switching devices. As mentioned in the first chapter, in electrostatically driven MEMS the pull-in voltage is a critical threshold of the imposed DC/AC bias, at which instability between electrostatic attraction force and mechanical restoring force take place [4]. As a result, at pull-in voltage the MEMS collapses abruptly on the underlying electrodes, thus establishing metal-to-metal contact between the input/output terminations, if the device is ohmic, or realizing the high-capacitance configuration, if the switch is designed to be capacitive.

The reference design is a quite standard configuration of MEMS movable membrane realized in the CMM-FBK RF-MEMS technology platform. It consists

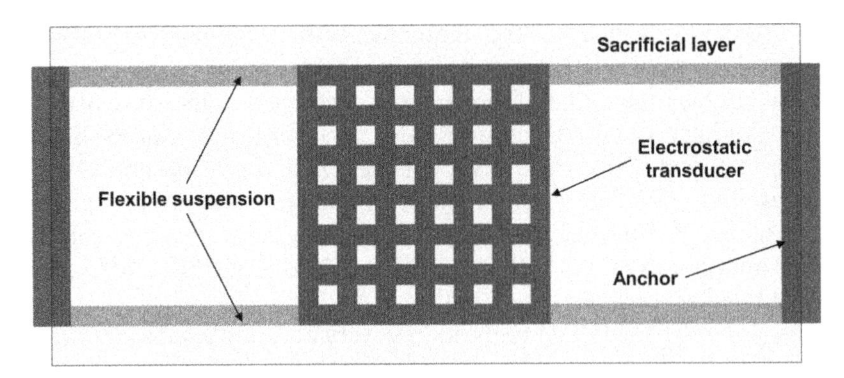

Figure 4.3. Layout of the clamped–clamped MEMS electrostatic micro-relay case study.

of a central square plate for electrostatic transduction, kept suspended above the underlying counter-electrode and input/output contacts by means of four deformable straight beams. Mechanical anchoring areas are placed at both ends of the suspended MEMS, giving rise to the so-called clamped–clamped configuration. The layout of the switch, realized in the L-Edit Layout Editor [5] is shown in figure 4.3.

The central electrostatic transducer is a 130 μm by 130 μm plate, joined to four 120 μm long and 10 μm wide flexible suspensions. The plate has 10 μm square holes (spaced at 10 μm from each other) necessary for the effective removal of the sacrificial layer underneath its surface [2]. Finally, external anchoring overlaps the sacrificial layer, thus realizing a gold vertical transition to the substrate underneath that ensures mechanical support to the elevated membrane (air-gap). With reference to table 4.1, the whole structure in figure 4.3 is realized with the first 2 μm thick gold metallization (light colour in figure 4.3). Furthermore, the central plate and anchoring areas also comprise the second 3 μm thick gold metallization, thus yielding a 5 μm thick membrane (dark colour in figure 4.3). This is done because the central transducer and the anchoring parts are meant to be mechanically stiff. However, flexible suspensions should be as deformable as possible in order to keep the pull-in voltage low. As for studying the pull-in/pull-out characteristic no other parts of the RF-MEMS device are necessary. In figure 4.3 all the other layers referring to buried metallizations are omitted (see figure 4.2 and table 4.1).

The layout generated in L-Edit is a 2D structure. However, in order to make the model suitable for FEM simulations, it has to be extended to a 3D structure. To do so, the layout in figure 4.3 is exported from L-Edit in Drawing Exchange Format (DXF) [6] and imported in the FreeCAD [7] environment, where a proper thickness is assigned to both layers through an extrude operation. The 3D model of the switch is shown in figure 4.4.

Once the RF-MEMS switch 3D model is ready, it is exported from FreeCAD in the so-called Standard for the Exchange of Product model data (STEP) format [8], and then imported within the ANSYS Workbench [9] simulation environment; this being the actual tool for carrying out FEM-based analysis. Within Workbench, mechanical properties are applied to the 3D structure according to the information reported in table 4.1. Suitable Boundary Conditions (BCs) are also imposed on the

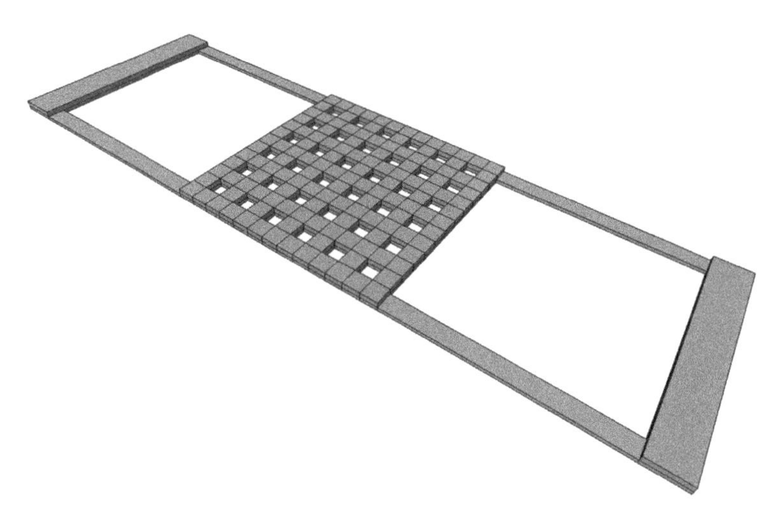

Figure 4.4. 3D model of the MEMS switch after extrusion with proper thickness of both layers performed in a FreeCAD environment.

model. In particular, fixed (zero) displacement is set for the anchoring areas, so that mechanical constraint operated by gold transition from the air-gap to substrate level is properly taken into account. On the other hand, electromechanical transduction is operated by including in the model TRANS126 elements through the Electro-Mechanical Transducers Generation (EMTGEN) macro [10]. In brief, a lumped parallel place capacitor element is generated per each node of the mesh underneath the transducer plate. An initial air-gap (rest position) of 3 μm is set with a minimum allowed gap of 100 nm. The latter is meant to avoid convergence issues when pull-in occurs, as the electrostatic force tends to excessive values when the gap is too small. As dynamic effects are not relevant in observing the pull-in/pull-out characteristic, DC static structural analysis is appropriate for this type of investigation. The voltage imposed across TRANS126 elements is ramped from 0 V up to 60 V, and then back to 0 V. The Workbench 3D schematic of the pulled-in (actuated) model is shown in figure 4.5, where the colour scale refers to the extent of vertical displacement. Deformation was magnified in order to make it visible. The way the 3D model was subdivided into mesh elements is also displayed.

The pull-in/pull-out characteristic, i.e. vertical displacement of the central plate versus applied DC bias, resulting from the simulation, is reported in figure 4.6. In detail, the pull-in takes place at 51.5 V, while the pull-out happens at 7 V.

Recalling what has been mentioned before concerning 5G specifications, the just-listed values are both quite high. Therefore, efforts at the design level should be invested with the aim of lowering the activation/deactivation DC levels. One of the Degrees of Freedom (DoFs) available is to reduce the elastic constant of flexible suspensions, e.g. increasing their length. A possible solution to reach this target while keeping the RF-MEMS switch design compact, is through designing folded (or meander-/serpentine-like) suspensions [11–12]. In light of these considerations, the modified layout in figure 4.7 is proposed.

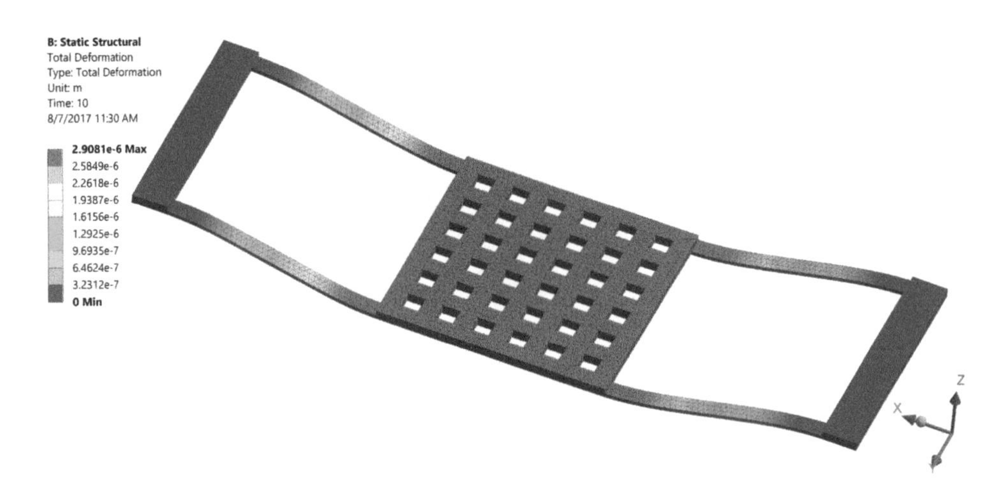

Figure 4.5. Workbench 3D schematic of the pulled-in RF-MEMS switch. The colour scale refers to the extent of vertical deformation. The displacement was magnified to make it easily detectable.

The 2D layout in figure 4.7 is exported/imported following the same procedure previously described for the structure in figure 4.3, and then simulated within Workbench. The pull-in/pull-out characteristic resulting from DC static structural simulation is reported in figure 4.8, showing activation and deactivation voltages of 24 V and 1.6 V, respectively.

With respect to the original layout with straight beams in figure 4.3, the pull-in and pull-out levels of the solution with folded-suspensions are lowered by 47% and 23%, respectively. Of course, the approach of serpentine-shaped suspensions can be further pursued, in order to keep on lowering the pull-in/pull-out threshold levels.

4.3.2 Transient dynamic simulation of the switching (opening/closing) time

Still bearing in mind the specifications derived by 5G requirements mentioned in the beginning of the chapter, another relevant performance indicator for RF-MEMS micro-relays is the switching time, as it is correlated to the latency of the system. To this purpose, transient structural analysis is performed in Workbench on both MEMS switch layouts previously reported in figure 4.3 and figure 4.7. The electrostatic transduction is implemented once again relying on TRANS126 elements. A 0–60 V square pulse is set as stimulus and is reported in figure 4.9(a).

The analysis is performed over a time interval of 2 ms, with 250 ns step resolution. Rise and fall times are 1 ns and take place at 100 μs and 500 μs, respectively (pulse length 400 μs; 20% duty-cycle). The response in terms of vertical displacement versus time is reported in figure 4.9 (b). Differing from static pull-in/pull-out characteristics, the rest position level on the vertical axis is set to 0 μm, in order to help visualize and assess overshooting during ringing at release. The pull-in transition is quite steep for both MEMS designs, while after the bias fall edge, different dynamic characteristics are evident. First, it should be highlighted that after the bias is zeroed, dynamic evolution of the MEMS structure is free (no stimuli applied). Therefore, its observation provides meaningful insight of the switch mechanical characteristics.

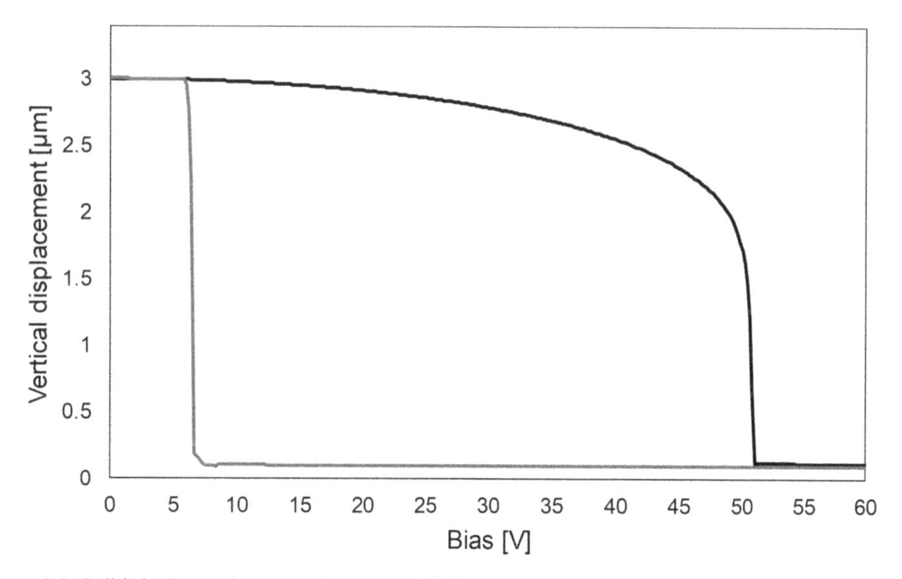

Figure 4.6. Pull-in/pull-out characteristic of the MEMS switch design in figure 4.3 resulting from DC static structural analysis performed within Workbench.

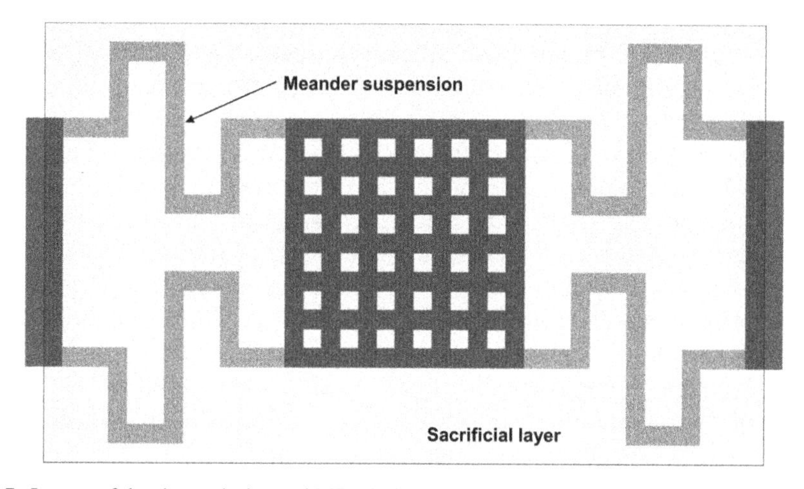

Figure 4.7. Layout of the clamped–clamped MEMS electrostatic micro-relay with meander-like folded flexible suspensions.

In particular, the frequency of damped ringing, visible in figure 4.9 (b), is determined by the elastic constant of flexible suspensions, and corresponds to the fundamental resonant mode of the mass-spring system composed by the rigid central plate and deformable beams [13]. Looking at the oscillations of the MEMS plate, highlighted in the close-up reported in figure 4.10 (b), the period measured for the MEMS with straight beams and folded suspensions is of 51 µs and 84 µs, respectively, corresponding to the resonant frequency of 19.6 kHz for the former and 11.9 kHz for the latter. Modal eigenfrequency analysis of both structures was performed in

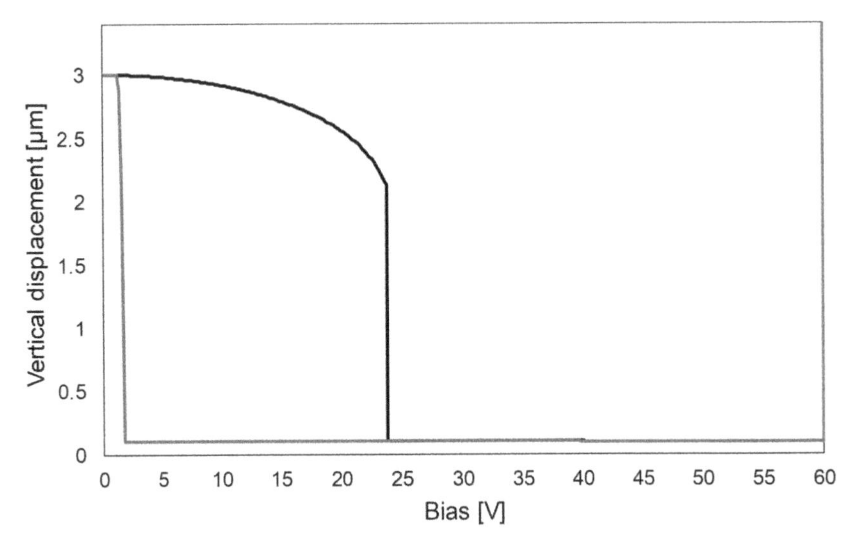

Figure 4.8. Pull-in/pull-out characteristic of the MEMS switch design in figure 4.7 resulting from the DC static structural analysis performed within Workbench.

Workbench [14], albeit not reported here for brevity, with the aim to cross-check these values. The results obtained concerning the fundamental resonant mode frequency of the case studies with different suspensions geometry are in close agreement with the numbers reported above, thus confirming the soundness of the modelling approach.

The air damping effect is not explicitly accounted for in this set of simulations, in order to probe into the dynamic of the MEMS structure itself. Nonetheless, a cumulative damping factor can be set in the transient simulation, to introduce dissipative energy effects that model the main losses of physical systems, like the damping intrinsic to the material (thermoelastic damping) [15–17] and, in fact, dissipative phenomena due to the viscosity of air. In the simulations reported here, the cumulative damping factor is set to a quite low and conservative value, in order to weakly account for viscosity of air and other dissipative effects. In any case, modelling of viscous damping due to the gas into which the MEMS device is immersed, was previously carried out, and it can be fully embodied in the simulations, also accounting for the squeeze damping effect [2–18] and gas flow across holes of the perforated central transducer [19–21]. Given these statements, damping effects induce a more pronounced attenuation of ringing amplitude in the evolution of the MEMS with folded suspensions (figure 4.7) rather than with straight beams (figure 4.3). This happens because the energy stored by the MEMS device in figure 4.3, when kept actuated, is larger due to greater spring constant of flexible suspensions [4].

Further considerations are now going to be developed around the switching (closing/opening) times for both the MEMS geometries discussed. To this purpose, close-ups of the dynamic evolution previously shown in figure 4.9 (b) are reported for the structures actuation and release in figure 4.10 (a) and figure 4.10 (b),

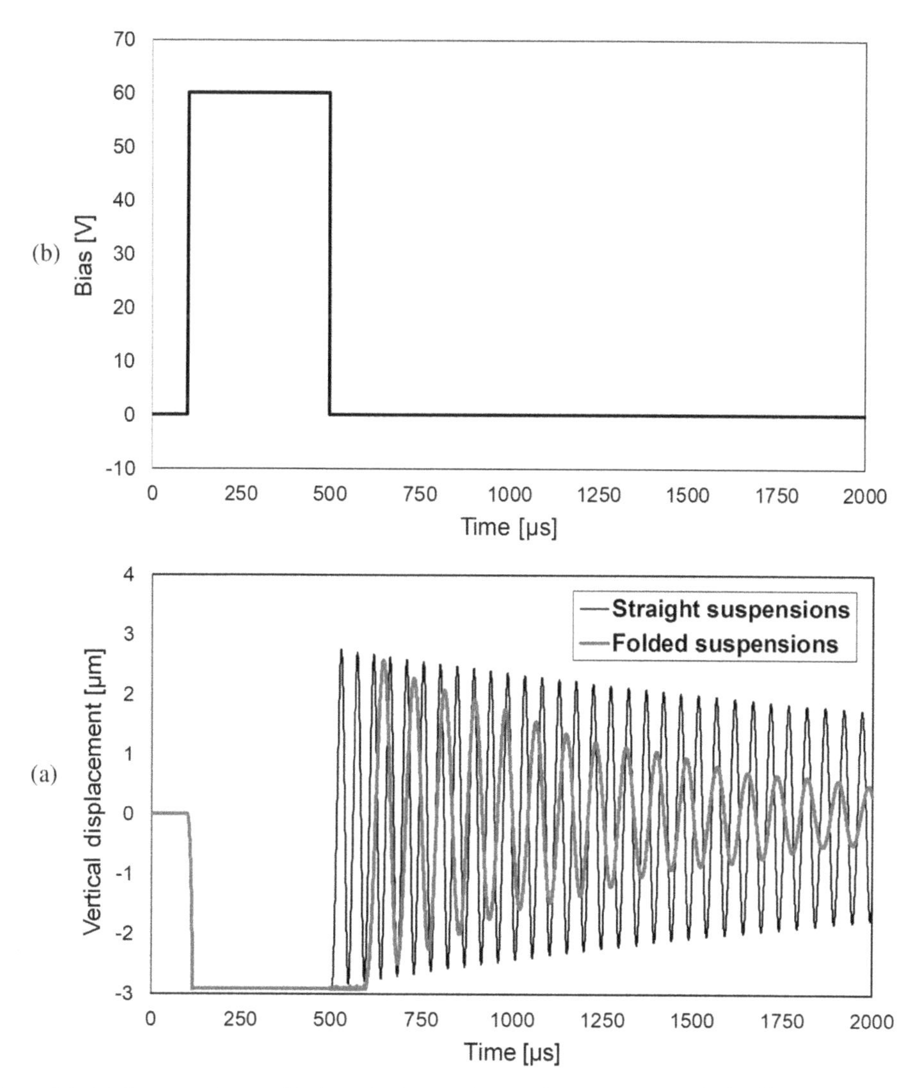

Figure 4.9. (a) Square pulse provided as stimulus in the transient structural analysis. (b) Dynamic vertical displacement response of the MEMS switch layouts discussed in figure 4.3 and figure 4.7.

respectively. In both plots, the origin of the time axis is placed at the proper edge of the square pulse depicted in figure 4.9 (a), i.e. rise edge at 100 µs, and fall edge at 500 µs.

Focusing at first on the actuation, the plot in figure 4.10 (a) shows that the characteristics of the MEMS devices are very similar to each other. Contact to the underlying plane takes place around 18 µs after bias commutes from 0 V to 60 V. Small ripples are visible for a couple of µs after the contact is reached. However, they are too small to be attributed to after-impact mechanical bouncing [22–23]. More likely, such small fluctuations are due to numerical convergence stabilization of the simulator in dealing with abrupt discontinuity of the physical contact that does not

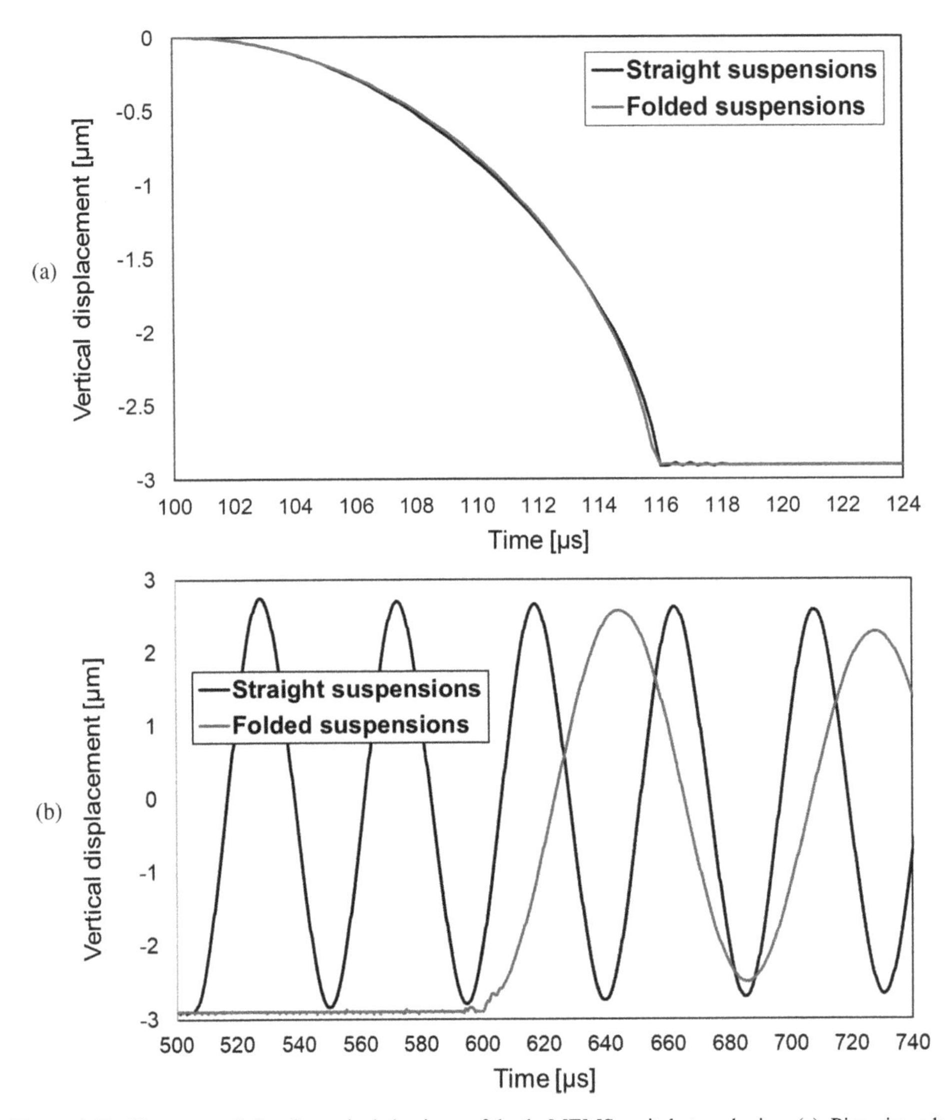

Figure 4.10. Close-ups of the dynamic behaviour of both MEMS switch topologies. (a) Bias rise edge (actuation). (b) Bias fall edge (release).

take place smoothly, but rather suddenly. In any case, steady pulled-in (actuated) configuration is established in less than 18 μs.

However, the release phase, reported in figure 4.10 (b), highlights a different behaviour of the two MEMS geometries. The switch with straight suspensions starts detaching from the underlying contact surface around 6 μs after the square pulse fall edge. Nevertheless, the design variation with folded suspensions commences moving upward around 100 μs after the bias drops from 60 V to 0 V. A longer release time can reasonably be expected for mechanical structures with lower elastic constant of suspensions, such as in the case of meander-like designs. As a matter of fact, lower

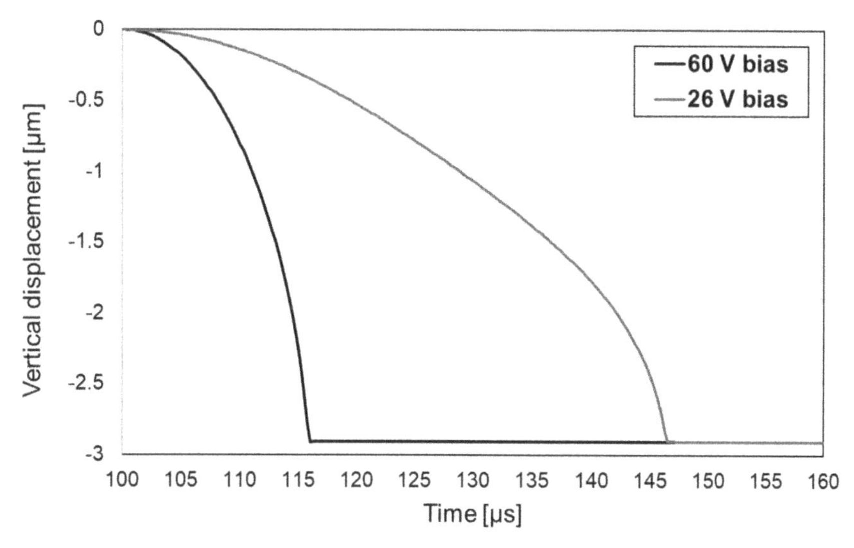

Figure 4.11. Dynamic behaviour of the MEMS switch with meandered suspensions at bias rise edge. (a) Bias peak value of 60 V. (b) Bias peak value of 26 V.

spring constant leads to less restoring force when the MEMS is actuated and, in turn, to slower release dynamics.

The considerations just developed highlight the existence of conflicting specifications, depicting the typical context within which MEMS designers have to come up with effective trade-offs. If, on one hand, lowering the elastic constant enables significant reduction of the pull-in voltage, on the other hand it makes the device dynamic response slower. The latter statement holds validity for the actuation phase as well. One should not be misled by the plot in figure 4.10 (a), where the transition time is the same for both MEMS designs. In that case, 60 V was imposed on the switch with meandered suspensions, while its pull-in voltage is around 24 V, as reported in figure 4.8. To this purpose, the plot in figure 4.11 shows the meandered switch dynamic for different bias levels.

Lowering the square pulse peak value from 60 V to 26 V, i.e. slightly above pull-in, transition time steps from 16–18 µs to 47 µs. In addition, reducing the MEMS device restoring force makes it more prone to failure mechanisms, like charge accumulation [24–25], micro-welding [26–27], and so on, that can lead to stiction, i.e. missed de-actuation when the bias is zeroed.

In conclusion, wrapping up the indications observed up to now, the proposed MEMS switch case study designs scored transition times better than 20 µs for actuation, and better than 100 µs for release, with pull-in voltage around 25 V. Nonetheless, conflicting trends among different specifications emerged as well, requiring careful investigation in order to identify optimum trade-offs.

4.3.3 Compact analytical modelling of RF-MEMS

Another approach to multiphysics simulation and analysis of microsystems that is worth mentioning is that of compact (analytical) modelling. Differing from FEM tools, where the 2D/3D model is decomposed in sub-elements and constitutive equations are solved for each of them, compact models are much simpler from the point of view of computation. Simplified descriptive analytical formulation is determined for the structure to be analysed, incorporating only the fundamental physical effects meant to be observed. Mathematical models can also be complemented by using empirical formulations, extracted for instance from previously collected experimental datasets. Bearing in mind equations (1.3) and (1.4) in chapter 1 that describe analytically the pull-in and pull-out of a parallel plate capacitor, their implementation in a programming language for automated computer-based resolution is already an example of compact modelling.

Of course, simplification of the analytical description brings a settlement of advantages and disadvantages. Concerning the former, compact modelling enables fast and computationally light simulations, thus considerably easing parametric analyses involving several DoFs. Furthermore, this agility makes it possible to incorporate more physical domains into the analysis. This enables sub-system/system level simulations, where the RF-MEMS electromechanical and electromagnetic behaviours are described together with that of CMOS circuitry. On the other hand, the accuracy of simplified models is, in general, not as pronounced as that of FEM tools, albeit satisfactory in most part of the cases RF-MEMS designers have to deal with. Also, importantly, the range of validity of simplified models is, by definition, quite limited. Still recalling the example of equations (1.3) and (1.4) in chapter 1, they describe the case of a parallel plate electrostatic transducer. If the MEMS movable electrode is not parallel but tilted with respect to the fixed counter-electrode, the model is not valid anymore. Similarly, if one wants to account for the fringing effect of the electric field at the borders, or for the presence of holes, or, again, for non-planarity and roughness of surfaces into contact, the model should be significantly modified/extended.

An effective solution to deal with the specificity and the limited validity range of compact models is to make use of modularity. Some compact models based simulation tools for MEMS and RF-MEMS provide a library of elementary components, which can be assembled together at a schematic level to compose the desired device geometry to be investigated. Basic elements are, for instance, straight flexible beams, parallel/non-parallel plate transducers, mechanical anchoring points, metal-to-metal contacts, and so on. The implementation discussed in [2–18] features a library of compact models developed by means of the VerilogA Hardware Description Language (HDL) [28] within the Cadence Integrated Circuit (IC) development framework [29]. An example of how this tool is utilized is shown in figure 4.12. The studied device is a cantilever RF-MEMS series ohmic switch belonging to a reconfigurable RF power attenuator [30–31]. Figure 4.12 (a) shows the correspondence between the device 3D topology, obtained with a white light interferometer (colour scale corresponds to the height along the vertical axis), and the Cadence schematic of the same device, composed of the basic elements available in the library. The pull-in/pull-out characteristic of the schematic model in figure 4.12 (a) is simulated by the Spectre Circuit Simulator [32] via static DC analysis. It is also observed experimentally by means of an interferometry-based

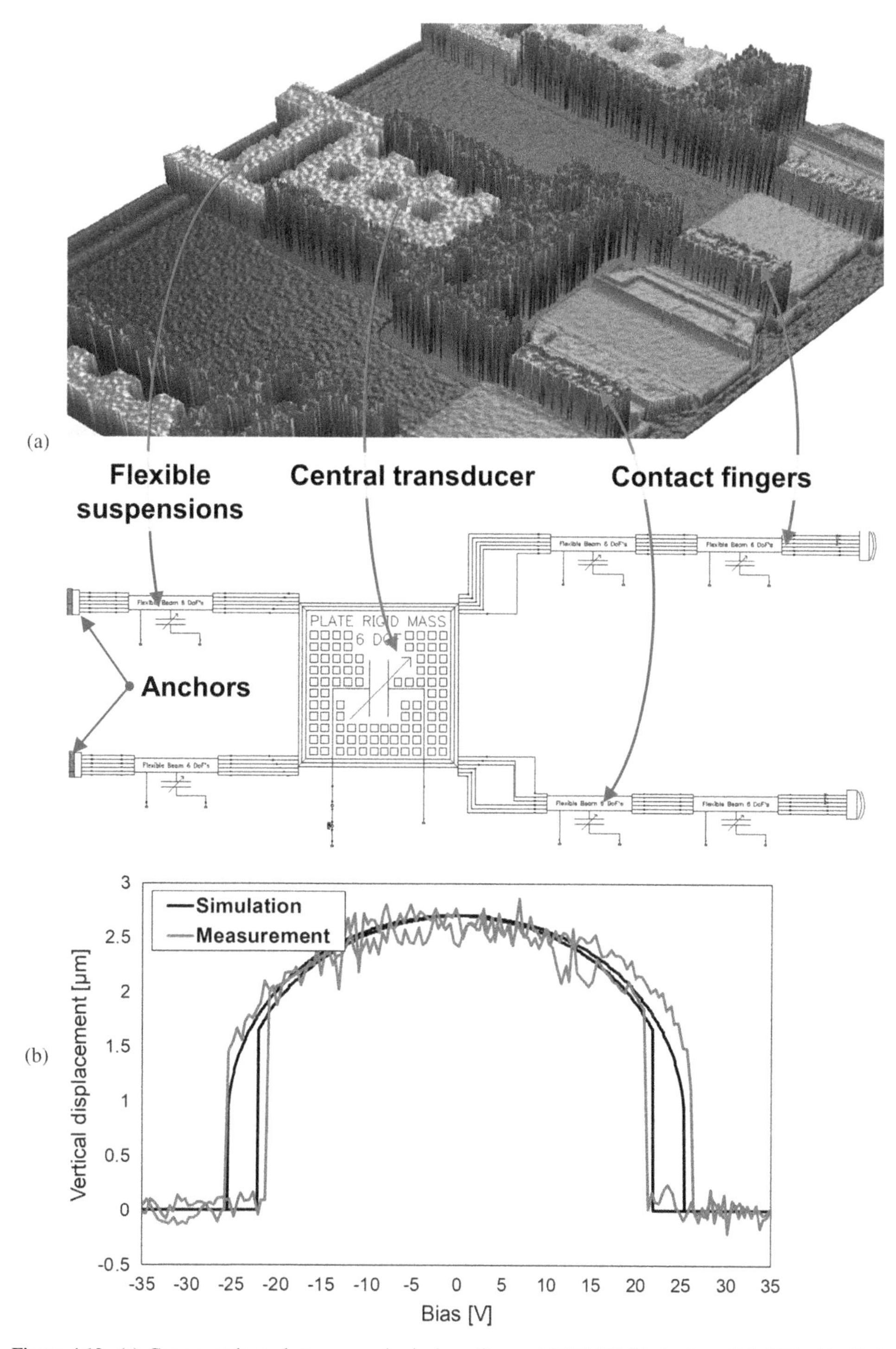

Figure 4.12. (a) Correspondence between a physical cantilevered RF-MEMS ohmic switch (3D white light interferometry) and the schematic built upon elementary compact models included in the library discussed in [2–18]. (b) Comparison of the pull-in/pull-out characteristic simulated by Spectre in the Cadence environment and the experimental one measured by pulsed light interferometry.

3D profiling system with pulsed light. The latter operation mode enables the measurement of Devices Under Test (DUTs) in dynamic conditions, as long as their behaviour is periodic [33–34]. The comparison of the simulated and measured actuation/release characteristic is reported in figure 4.12 (b), where the accuracy of the model in predicting the pull-in and pull-out levels, at 26 V and 22 V, respectively, is visible. Following the same approach, it is possible to extend the compact model in order to incorporate the RF behaviour as well, while keeping the computational complexity quite moderate.

The compact model library reported above is not the only example of compact modelling applied to RF-MEMS. Valuable contributions from other researchers are discussed and reported in the literature [35–40], covering a wide range of modelled behavioural features, such as, for instance, inertia, residual stress of metals, as well as squeeze film damping.

Based upon approaches similar to that discussed in this subsection, open source simulation software packages as well as commercial products have been released, like the SUGAR simulation tool [41–42], the NODAS (Nodal Design of Actuators and Sensors) [43–48] and the ARCHITECT by Coventor [49].

4.4 Electromagnetic simulation of RF-MEMS devices

The focus of this section is the modelling and simulation of RF-MEMS electromagnetic characteristics. Following an approach similar to the one previously developed when dealing with electromechanical behaviour of microsystems, both FEM analysis and extraction of equivalent lumped element networks are going to be reviewed and discussed.

4.4.1 Finite Element Method (FEM) analysis

Among the available commercial software tools appropriate for conducting FEM electromagnetic analysis of RF-MEMS, the Ansys HFSS (High Frequency Structural Simulator) [50] is selected and exploited in the following examples.

Validation of the simulation tool

Prior to exploitation of HFSS for design refinement versus target specifications, it is fundamental to build confidence on the accuracy of the simulated results. As a matter of fact, besides the appropriateness of a certain simulator in solving the problem of interest, several settings must be determined in the proper way if accurate, credible and repeatable results are to be achieved. Moreover, reasonable trade-offs must also be scored concerning the accuracy of results and computation complexity. For instance, if an accuracy improvement of 0.3–0.5% means increasing the simulation time from 20–30 min to 5–7 h, it is pointless to try.

All these aspects must be solved by carrying out thorough validation of the simulation tool, having a population of a few significant RF-MEMS design concepts and experimental datasets; in this case scattering parameter (S-parameters) measurements, collected from the physical samples available. In order to perform a validation that is sound and robust to a fair extent, DUTs of different complexity should be selected, such as simple test structures as well as multi-state high-order devices, and the physical specimens should come from different wafers and fabrication batches.

When a certain simulation tool is employed for the first time, a reasonable choice is to start from the nominal technology values, for what concerns the electrical properties of each material as well as for the geometry features, such as layer thickness, planarity, roughness, and so on. Then, by confronting the results of simulations against measurements, the post-layout verification phase (already mentioned in section 2.4.1 and reported in figure 2.6) is performed, leading to a set of actual values describing the specific technology and its spreads. Of course, simulations versus measurements disagreement due to technology non-idealities must not be mixed or confused with that due to the settings of the simulator itself. In fact, several simulator features can influence the quality of results. Among them stand, for example, the mesh type and density, the excitations providing RF power to the 3D structure, the selected Boundary Conditions (BCs), the solver algorithm, and so on.

An effective approach to separate technology-driven inaccuracies from those caused by intrinsic simulator settings, is to start the validation with nearly-ideal devices. For example, if simple Coplanar Waveguide (CPW) and/or microstrip structures are available, it is reasonable to assume that their S-parameters' behaviour is quite close to ideality. Therefore, benchmarking simulations against such experimental datasets helps determine the most appropriate simulator settings. In addition, or as replacement of physical CPW/microstrip test structures, the simulator can be tuned upon data produced by other (already validated) simulation tools, and/or by trusted analytical models.

For the purposes of this section, the validation of the HFSS FEM tool against S-parameters' measurements is carried out for the RF-MEMS series ohmic switch, whose layout is discussed in [51–52] and depicted in figure 4.13. In order to simplify the understanding of the proposed geometry, only metal layers defining the elevated MEMS membrane and the sacrificial layer are shown. However, the buried layers, biasing lines, surrounding CPW and protective package [53] are hidden, despite being present in the L-Edit layout file.

Pursuing a similar approach to the one discussed in the section dedicated to electromechanical simulation of RF-MEMS, starting from the layout in figure 4.13, a full 3D model of the RF-MEMS switch is generated (including layers and structures not depicted above) and then imported into the HFSS graphic environment. Material properties and thickness of layers are set according to the information previously reported in table 4.1. HFSS simulation settings are defined as discussed in detail in [54] and not reported here for the sake of brevity. The full 3D model of the RF-MEMS switch imported into the HFSS design environment is shown in figure 4.14. The vertical position of the MEMS elevated membrane was parameterized, in order to easily adapt the 3D model to the ON and OFF switch configurations.

On the experimental side, the S-parameters are characterized on a probe station (on-wafer measurements) by means of a Programmable Network Analyser (PNA) [55]. RF-MEMS physical samples are contacted with Ground-Signal-Ground

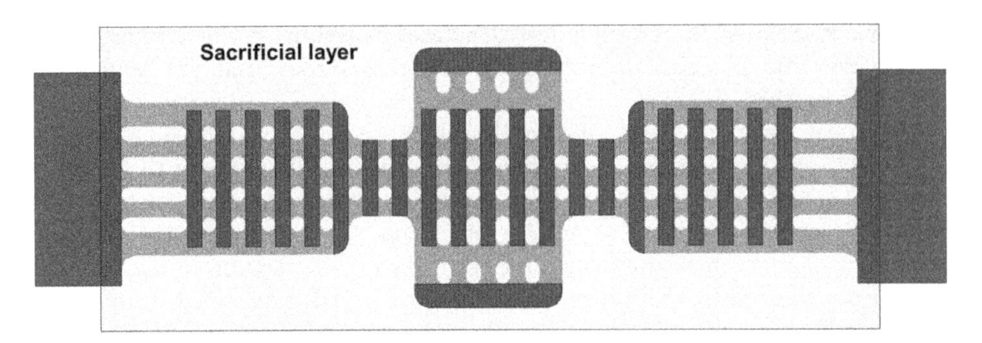

Figure 4.13. Layout of the RF-MEMS series ohmic switch exploited for validation purposes. The sacrificial layer, defining the elevated membrane, is visible.

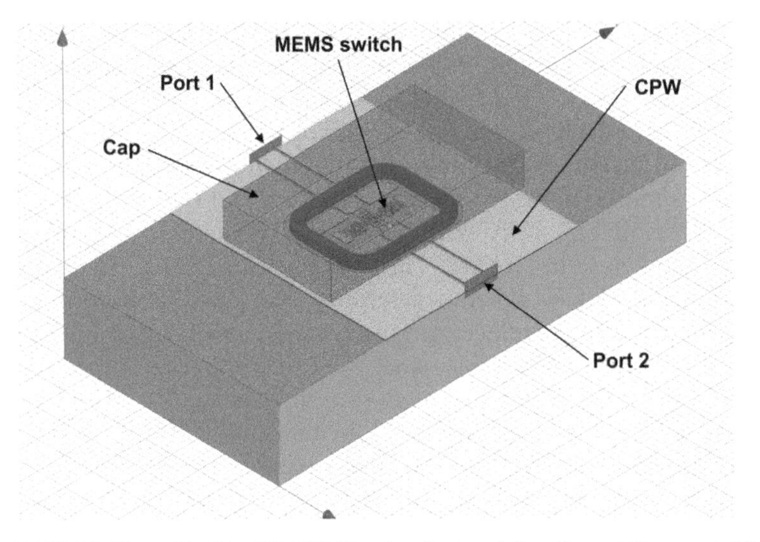

Figure 4.14. HFSS full 3D model of the RF-MEMS series ohmic switch in figure 4.13 generated from the 2D layout. It includes all the layers of the CMM-FBK technology, the CPW surrounding frame and the quartz protective cap.

(GSG) probes (150 μm pitch). Delivered power is 0 dBm and the chosen calibration routine is the Line-Reflect-Reflect-Match (LRRM), performed up to 110 GHz. In order to obtain better accuracy, LRRM calibration and subsequent measurements were performed in two distinct frequency ranges, namely, from 10 MHz to 67 GHz (Low Frequency Range—LFR) and from 67 GHz to 110 GHz (High Frequency Range—HFR). A photograph of a few RF-MEMS physical samples placed on the probe station for RF measurements is shown in figure 4.15. GSG probes for S-parameters characterization as well as DC probes to provide biasing for the actuation of the elevated MEMS membranes are visible.

The comparison of HFSS simulations against S-parameters' measurements are reported in figure 4.16 and figure 4.17 for the RF-MEMS switch in OPEN (OFF state) and CLOSE (ON state) configuration, respectively. Concerning the latter, the MEMS elevated membrane was pulled-in by applying a DC bias level of 50 V. In more detail,

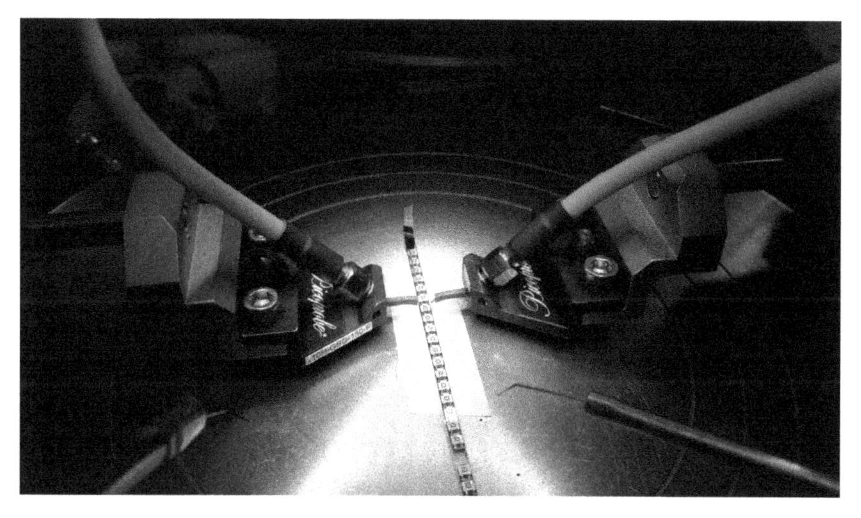

Figure 4.15. Photograph of RF-MEMS series ohmic switch physical samples (see figure 4.13 and figure 4.14) accommodated on the probe station for S-parameters' measurements.

figure 4.16 (a) shows the reflection parameter (S11) and figure 4.16 (b) the input/output isolation (S21) of the OPEN micro-relay. Figure 4.17 (a) reports the reflection parameter (S11) and figure 4.17 (b) the loss (S21) of the CLOSE switch, respectively.

Recalling the validation purposes discussed above, the results just mentioned highlight a very good accuracy of the simulated traces against experimental measurements, thus proving that the 3D model building procedure, the assigned effective material properties, as well as the simulator settings, are properly chosen and defined. The methodology's robustness is also proven by the fact that the accuracy of simulations does not get worse when the state of the DUT is commuted from the isolating (OFF state) to conducting (ON state) configuration. In addition, it should also be borne in mind that the analysed frequency range is very wide, as it spans from 10 MHz up to 110 GHz. These considerations further support the broad usability of the discussed simulation approach, both concerning RF-MEMS structures with different geometries/configurations, as well as diverse frequency ranges of interest.

Transcending the accomplished validation task, the results reported in figure 4.16 and figure 4.17 offer an effective insight in order to start a discussion on the suitability of RF-MEMS devices for 5G applications. To this purpose, it is useful to refer to the set of target specifications discussed at the end of section 4.1, with particular reference to isolation and loss.

Looking at the plot in figure 4.16 (b), isolation (S21 when the switch is OFF) is better than −30 dB up to 14 GHz, and better than −20 dB up to 40 GHz. Nonetheless, between 40 GHz and 110 GHz isolation is never better than −20 dB, and the worst value it scores (in the simulated trace) is −10 dB around 62 GHz. However, in agreement with the target specification discussed in section 4.1, isolation should be better than -30/-40 dB, up to the higher frequency possible. According to that, the RF-MEMS switch design reported in figure 4.13 is able to

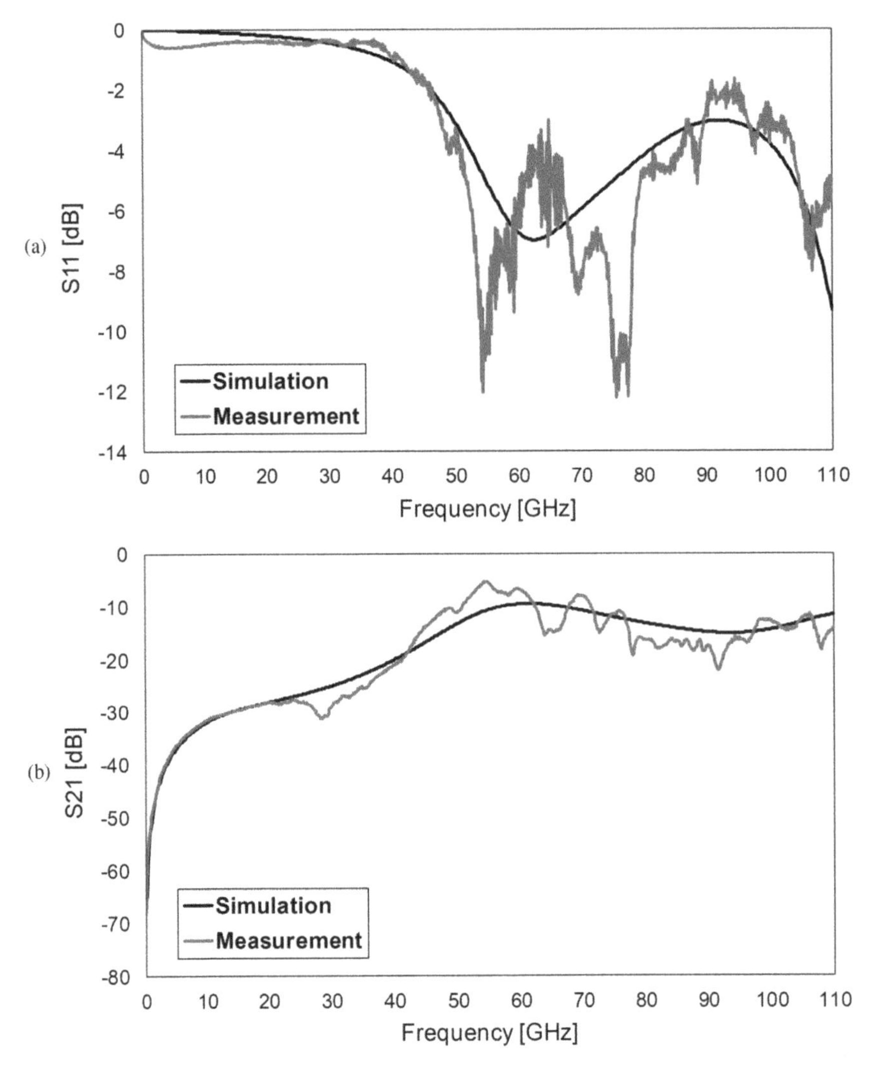

Figure 4.16. Simulated versus measured S-parameters' characteristics of the RF-MEMS device in figure 4.13 from 10 MHz up to 110 GHz when the DUT is OFF (switch OPEN; MEMS in the rest position). (a) Reflection (S11 parameter). (b) Isolation (S21 parameter).

satisfy the requirement on isolation just up to 14 GHz. Therefore, effort must be invested with the aim of improving the switching unit design.

Similarly, figure 4.17 (b) provides meaningful insight concerning the RF-MEMS micro-relay loss while conducting (S21 when the switch is ON). In the discussion developed in section 4.1, loss should be ideally better than −1 dB, up to the highest frequency possible. According to the simulated trace in figure 4.17 (b), loss is better than −1 dB up to 24 GHz, better than −2.6 dB up to 55 GHz, and better than −4.5 dB up to 80 GHz. Above the latter frequency, however, the S21 worsens quite steeply, reaching a poorest value of about −18 dB at 110 GHz.

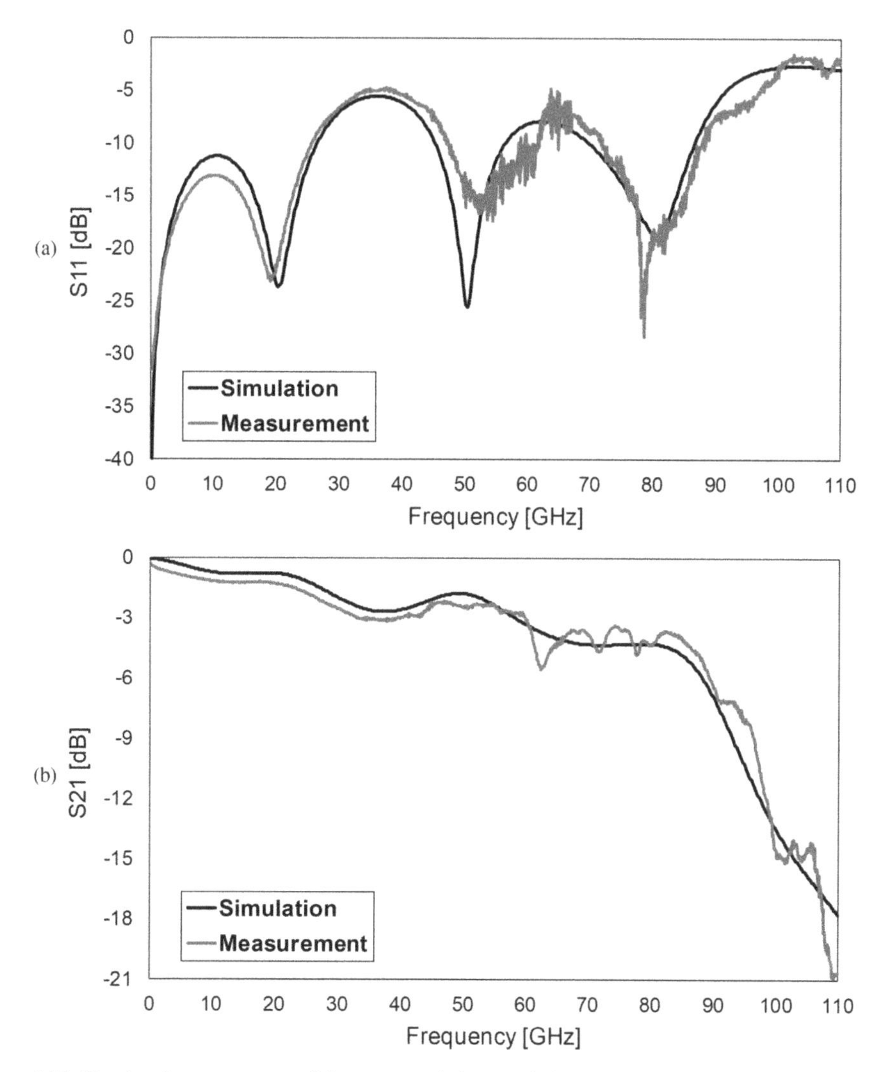

Figure 4.17. Simulated versus measured S-parameters' characteristics of the RF-MEMS device in figure 4.13 from 10 MHz up to 110 GHz when the DUT is ON (switch CLOSE; pulled-in MEMS). (a) Reflection (S11 parameter). (b) Loss (S21 parameter).

In conclusion, although the RF-MEMS switch topology reported in figure 4.13 is not meant for 5G applications, but rather reported here for validation purposes, it exhibits promising characteristics in terms of S-parameters. Therefore, it is a valuable case study to be employed as a starting point for further design optimizations.

Design variations and performance optimization aimed to 5G requirements

In this section, a few examples of RF-MEMS switch/switching unit design variations are going to be proposed, in order to investigate their impact on the S-parameters' characteristics, in light of the target specification discussed at the beginning of this chapter.

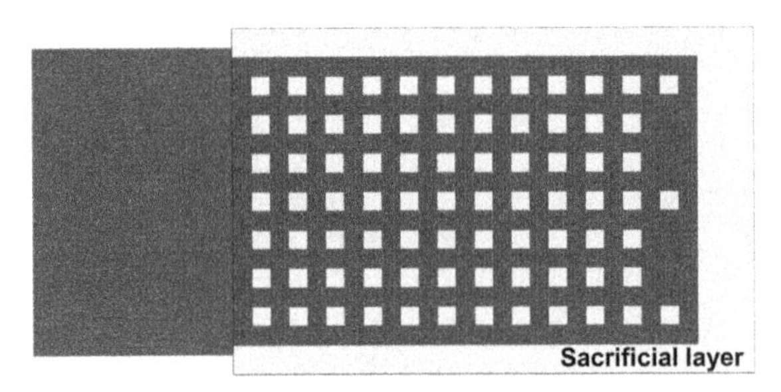

Figure 4.18. Layout of the RF-MEMS series ohmic cantilever-type (i.e. single-hinged) switch.

At first, the switching unit clamped–clamped topology in figure 4.13 is replaced by a cantilever-type geometry, such as the one reported in figure 4.18.

The suspended membrane is hinged only on one side, while the other end of the elevated gold layer is free. Underneath it, metal contacts (not shown in figure 4.18) are placed, thus ensuring input/output metal-to-metal continuity, when the MEMS membrane is pulled-in. Beside the substantial difference between the clamped–clamped and cantilever geometry just mentioned, other aspects must be highlighted. Concerning the micro-relay placement with respect to the CPW framing it, clamped–clamped geometries are typically deployed transversally with respect to the RF signal line. On the other hand, cantilever geometries can easily be accommodated in-line along the signal path, as in the case shown in figure 4.18. This leads to more compact arrangements of the switching device, without the need for recesses in the RF ground planes to allocate the MEMS membrane, as in the case of clamped–clamped configurations. Such characteristics suggest smaller influence of cantilever versus clamped–clamped switches on loss and mismatch of the micro-relay when ON (i.e. CLOSE switch) [56–58]. Concerning the electromechanical properties of cantilever-type micro-relays, having the membrane hinged only on one end, leads intrinsically to lower pull-in voltage levels, if compared to clamped–clamped geometries. Nonetheless, on the other side, the limited restoring force exposes cantilevered MEMS micro-relays to malfunctioning due to charge entrapment and micro-welding, making them, in general, less robust than clamped–clamped switches versus power handling and cycling [59–61].

The simulated S21 characteristics of the switch topologies in figure 4.13 and in figure 4.18 (isolation when OFF; loss when ON) are reported in figure 4.19. In order to have a consistent comparison of the two geometries, in the HFSS 3D model exclusively the intrinsic switch was modified, while the surrounding CPW structure was unaltered.

Probing in more detail, the plot in figure 4.19 (a) compares the isolation (S21) of the two geometries when both micro-switches are OFF. In general terms, the original clamped–clamped design solution in figure 4.13 performs better. The two characteristics are quite similar in the frequency range from 51 GHz to 65 GHz. On the other hand, from 200 MHz up to 51 GHz, isolation of the clamped–clamped micro-relay is always better, with a difference compared to the cantilever that reaches a maximum of about 10 dB around 20 GHz. From 65 GHz up to 110 GHz

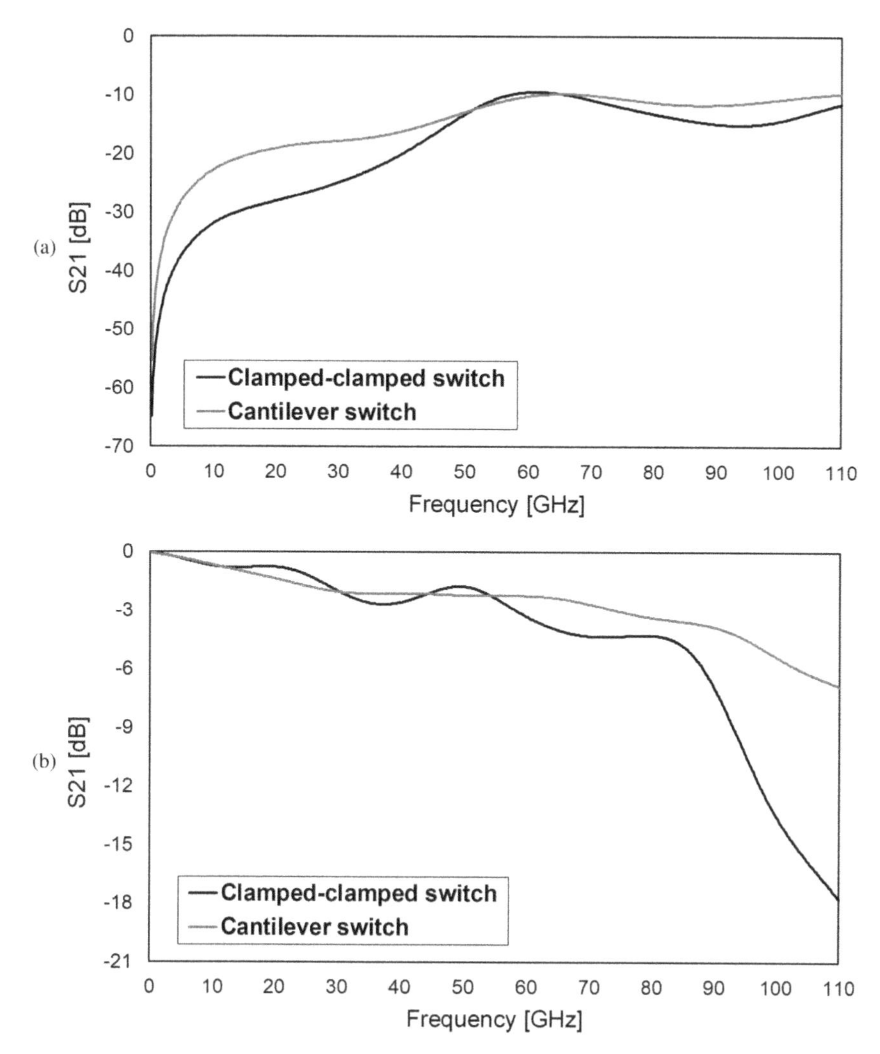

Figure 4.19. Simulated S-parameters characteristics of the clamped–clamped and cantilever series switches, depicted in figure 4.13 and in figure 4.18, respectively, from 200 MHz up to 110 GHz. (a) Isolation (S21 parameter) when micro-relays are OFF (i.e. OPEN switches). (b) Loss (S21 parameter) when micro-relays are ON (i.e. CLOSE switches).

the isolation divide between the two geometries is less pronounced. In any case, the geometry in figure 4.13 scores values better than a few dB with respect to the design solution in figure 4.18.

The fact that the clamped–clamped configuration outperforms the cantilever switch in terms of isolation can be explained in quite a straightforward fashion. The double-hinged switch has two contact areas above the RF line. When ON, such areas establish metal-to-metal contact ensuring the RF signal flow across the device. However, when OFF, overlapped contact pads realize two small parasitic capacitances placed in series configuration, which couple the input and output branches.

On the other hand, the cantilever membrane has just one contact area, which means a unique parasitic capacitance when OPEN. Since the contacts' area was kept the same in the two geometries, the clamped–clamped switch is characterized by a parasitic OFF state capacitance that is half of the one realized by the cantilever (two equal capacitors in series), thus ensuring better isolation. This issue can be mitigated by reducing the contact area in the cantilever design, albeit influencing the switch loss when ON.

Stepping forth, figure 4.19 (b) compares the loss introduced by the clamped–clamped and cantilever switches, when conducting (ON state), indeed overturning the situation described above. Considering the observed frequency range as a whole, the loss introduced by the cantilever is better than that of the clamped–clamped micro-relay. In more detail, the two curves are nearly identical up to 12 GHz. In between 12 GHz and 31 GHz, the cantilever performance is slightly worse than in the case of the double-hinged device, being in any case better than −2 dB. Nonetheless, between 31 GHz and 64 GHz the cantilever loss characteristic exhibits pronounced flatness, worsening from −2 dB to −2.3 dB from the lower to the higher frequency previously indicated. In the same range, the clamped–clamped loss oscillates between −2.7 dB and −1.8 dB up to 50 GHz, and then degrades to −4 dB at 64 GHz. Again, between 64 GHz and 82 GHz the cantilever loss is always better than −3.5 dB, while the clamped–clamped loss is better than −4.4 dB. Finally, from 82 GHz up to 110 GHz both characteristics become more significantly lossy. However, while the cantilever S21 trace is always better than −7 dB, the clamped–clamped loss reaches −18 dB at 110 GHz.

This solution has the unique advantage of modifying the switching unit intrinsic topology in order to improve device performance. Such a strategy leads unavoidably to dead ends in terms of conflicting characteristics optimization. To this regard, the example of the clamped–clamped versus cantilever switch geometry, comprehensively reported above, is quite significant. Having just one contact area (cantilever switch) minimizes the series ON state resistance, which leads to reduced loss when compared against devices with double contact areas (double-hinged switch). Nonetheless, a single contact area means doubling the OFF state capacitance that, in turn, brings worse isolation of the cantilever against the clamped–clamped geometry. Given the background just depicted, the only trade-off possible is sacrificing isolation in the name of smaller loss, or vice versa.

On a completely different reference plane lie plenty of opportunities in terms of performance optimization, if one pursues the strategy of making the design more complex. Looking in this direction, a valuable case study is reported in the following. When an RF-MEMS switching unit topology is optimized against the desired performance figure in terms of loss when CLOSE, the isolation (when OPEN) can be increased by adding redundant switching units working as isolation boosters. A reference layout of this concept is shown in figure 4.20.

The actual series switching unit is preceded and followed by two micro-relays with the same geometry of the suspended MEMS membrane, albeit configured as shunt ohmic micro-relays. Shunt switches have a dual behaviour with respect to series ones, as they are CLOSE when the MEMS is OFF, and OPEN when the

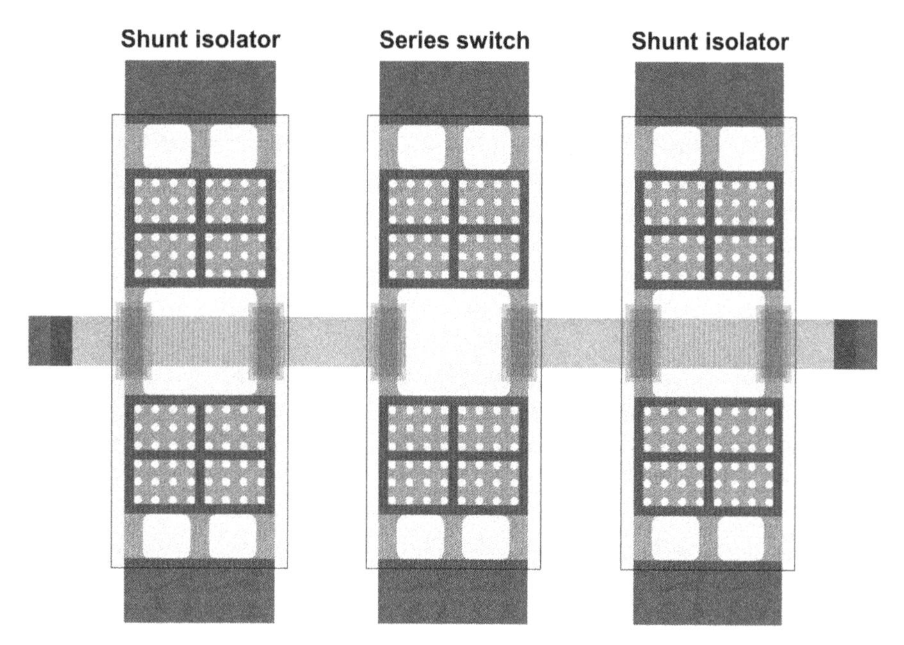

Figure 4.20. Layout of an RF-MEMS ohmic series switch with two isolation boosting units, placed before and after the intrinsic micro-relay, based upon shunt ohmic switches. The buried RF underpass is visible. It is continuous underneath the shunt units and interrupted underneath the series switch.

MEMS is ON. In the latter case, they divert (short) the signal to RF ground. Therefore, when the central micro-relay is OFF (OPEN switch), the two isolation boosters can be actuated (pulled-in) in order to short to ground the RF signal flowing across the intrinsic switch because of the parasitic capacitive coupling. This helps improve the isolation of the overall switching device in figure 4.20, as reported in figure 4.21.

When the isolation boosting units are not activated (OFF state), the only series RF-MEMS switch does not exhibit outstanding isolation (S21 parameter). Diversely, when the shunt ohmic micro-relays are actuated (ON state), the isolation shifts evidently towards improved performance. In detail, the S21 trace of the OPEN switch with isolation boosters ON in figure 4.21 scores enhanced values of 10 dB up to 25 dB when compared to isolation with boosters OFF, from 200 MHz up to around 84 GHz. Above the latter frequency and up to 110 GHz, the two curves are quite close to each other. Recalling the 5G reference specifications listed in section 4.1, the isolation of the design concept in figure 4.20 with isolation boosters activated reaches the target of being better than −30 dB from 200 MHz up to around 90 GHz.

Of course, the inclusion of isolation boosting units affects the loss performance of the switch when ON, as additional parasitic capacitance couples the shunt ohmic switches to RF ground when they are OFF. Therefore, increasing the geometry complexity does not rule out, by itself, the rising of conflicting characteristics. On the other hand, one should bear in mind that adding redundancy to the RF-MEMS design concept under development, introduces significant sets of DoFs the developer can play with. For example, still referring to the case in figure 4.20, one can choose

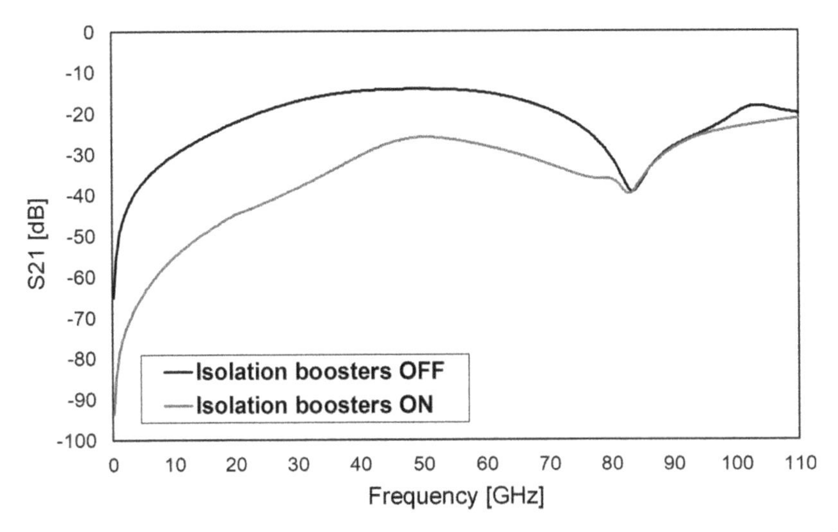

Figure 4.21. Simulated isolation (S21) parameter of the switching device shown in figure 4.20 when the isolation boosting units are OFF and ON, from 200 MHz up to 110 GHz.

the intrinsic switching unit tailoring its loss (when ON) characteristics to the best performance possible, without paying too much attention to its conflicting figure in terms of isolation, as the latter can be enhanced by introducing additional switching stages.

4.4.2 Modelling based on equivalent lumped element networks

The target of this concluding section is to provide an agile outlook of RF-MEMS modelling techniques based on equivalent lumped element networks. With similarity to the discussion previously developed around electromechanical modelling of MEMS, prediction of RF/electromagnetic behaviour can be performed following diverse approaches and by means of different tools. From a conceptual point of view, the approach to RF equivalent networks is very close to that of device behavioural description based on compact models, developed in section 4.3.3. As a matter of fact, the aim is to attain accurate prediction of RF-MEMS electromagnetic behaviour by keeping a limited computation load. Thereof, multiple DoFs design optimization runs can be performed quickly, while sub-system/system level simulations are also enabled. To this purpose, compact models describing electromechanical behaviour can be enriched by equivalent lumped element networks, thus allowing the prediction of the whole mixed-domain RF-MEMS behaviour in quite a straightforward manner.

Equivalent lumped element networks are widely used in RF/microwave modelling of electronic devices and circuits. Substantially, starting from experimental datasets (e.g. S-parameters), accurate simulated data (e.g. from already validated FEM models), as well as being based upon mathematical models, a network of capacitors, inductors, resistors and other components, is derived. Such an architecture is able to describe quite accurately the electromagnetic behaviour of the original device, within an established range of frequency. It should be highlighted that the three sources of data for the network extraction, i.e. measurements, simulations and analytical models, can be exploited in synergy, as well as exclusively. This means that, depending on the circumstances, a lumped element network of a certain device can be based, for instance, just on measured datasets, or on a part of experiments and of simulations, and so on. This suggests the pronounced flexibility of the discussed approach, which is also confirmed when probing into the scientific literature. To this regard, the variety of electronic components modelled by means of such a methodology is significantly wide. Several techniques and examples of lumped element networks describing semiconductor devices, transistors and circuits are available [62–64]. Furthermore, valuable contributions discussing equivalent networks of passives such as waveguides, inductors, transformers and so on, are also available [65–70]. Following the same path, modelling based on equivalent lumped element networks was also investigated in the field of RF-MEMS devices [11,71–75].

Selected case study and methodology

The example that is going to be discussed in the following is based upon elementary 1-bit (ON/OFF) RF-MEMS attenuator designs, previously discussed in [76–77]. The microphotographs of the devices for which the lumped element networks will be extracted, are reported in figure 4.22.

The RF-MEMS attenuator modules are available in two configurations, namely series and shunt, reported in figure 4.22 (a) and figure 4.22 (b), respectively. The RF-MEMS switch and surrounding CPW frame are unaltered in both variants. What changes is how the resistive load is deployed. In the series device, the polycrystalline silicon resistor is in-line on the RF central path, while in the shunt device two polycrystalline silicon resistors are placed in shunt-to-RF ground configuration. Therefore, in the series device in figure 4.22 (a) the RF signal is attenuated when the MEMS switch is OFF, while the load resistor is shorted when the MEMS switch is ON. However, in the shunt device in figure 4.22 (b) the RF signal is not attenuated when the MEMS switch is OFF, while the two shunt loading resistors are inserted (in parallel to one another) when the MEMS switch is ON.

The circumstance of design variations, in which for most part, the RF-MEMS device layout is kept fixed, while just limited portions are modified, such as in the case discussed here, is particularly favourable for extracting equivalent lumped element networks. To this purpose, a critical aspect must be carefully evaluated. Lumped networks able to mimic a measured/simulated curve, or the response of a mathematical model in a limited range of frequencies, can be arbitrarily defined. Nonetheless, this would not bring any added value to the analysis of RF-MEMS electromagnetic behaviour. However, what really makes sense is constructing lumped element networks topologies that capture the physical characteristics of the device under analysis, thus enabling a coherent description of it.

At this stage, a practical example can help one to understand the aforementioned critical aspect. Let us recall the basic equivalent network of a CPW, already discussed in chapter 1 and reported in figure 1.4 (c). The architecture makes sense

(a)

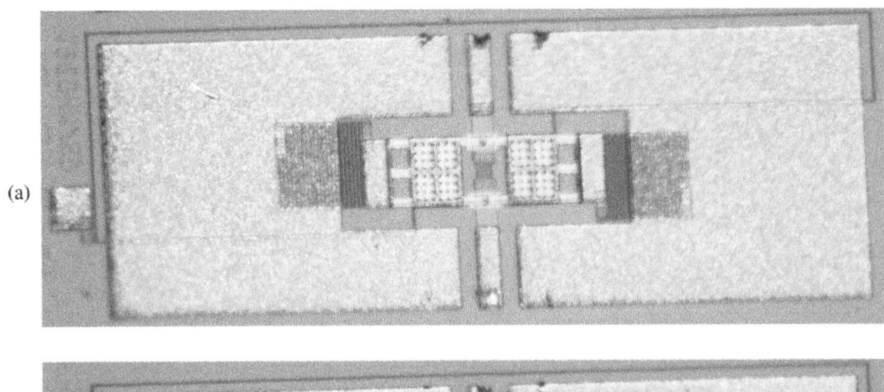

(b)

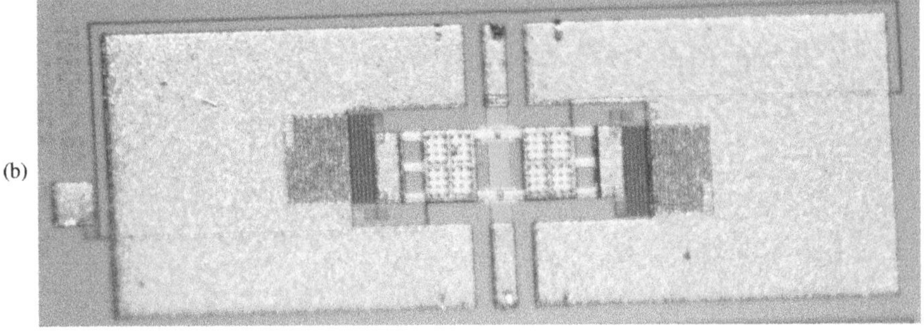

Figure 4.22. Microphotographs of the RF-MEMS 1-bit (ON/OFF) power attenuators discussed in [76–77]. (a) Series design, featuring one resistive load placed in-line on the RF path. (b) Shunt design, featuring two resistive loads placed in shunt-to-RF ground configuration.

from a physical point of view. The series resistance and inductance are those of the metal RF line, while the shunt capacitance and resistance model the capacitive coupling and losses to ground, respectively. If the CPW geometry was modified, it is quite straightforward to envisage how this would reflect on the topology in figure 1.4 (c). For instance, if the gap (between the RF line and the ground planes) is reduced, the capacitance to ground increases. If the substrate is replaced with a more lossy material, the resistance to ground will decrease (i.e. loss will increase). Similarly, if the CPW length is increased, both resistive and inductive series contributions will rise.

Following a different approach, various topologies, other than the one in figure 1.4 (c), can be put arbitrarily together to effectively reproduce the S-parameters characteristics of a CPW, also via automated computer-based extraction routines. Nonetheless, when a robust conceptual link between each lumped element and the physical feature/s it describes is missing, impairing limitations start to emerge. First, the network architecture does not make sense (partially or totally) from a physical point of view. Therefore, it is useless to develop a better understanding around the behaviour of the RF-MEMS under analysis. Then, the extracted network architecture is valid exclusively in the context within which it was derived. This means that if the geometry DoFs of the device are altered, as well as if the topology is modified (such as when stepping from the device in figure 4.22 (a) to that in figure

4.22 (b)), the lumped network can be neither parameterised nor partially reused/ported. On the other hand, it should be once again extracted from scratch.

This discussion clarifies the importance of building the equivalent lumped element network architecture, always sticking to sound physical considerations, and increasing its complexity in a smooth fashion, i.e. adding reactive/resistive elements always linked to real effects, and not just because they improve the match to target measured datasets.

For the same reason, once the starting network architecture is defined, the tentative value of each lumped element should be also carefully set. Such numbers should always be distilled starting from physical considerations, and not following random approaches. To this purpose, possessing sound knowledge both of how the device is structured and of the basic lumped elements' analytical models, gives a significant helping hand. For instance, if the developer, besides the 2D layout, knows also the thickness of the metal and insulating layers, he/she can determine quite confidently the value of the parasitic OFF state capacitance of an RF-MEMS. Similarly, if the series inductance contribution due to a portion of the RF line has to be determined, the developer is prompted to calculate the inductance per unit length, on the basis of the line dimensions and material properties. Therefore, it will be easy to derive a value making physical sense.

Once the initial values of each lumped element are determined following the approach just discussed, they can be tuned within a limited range of variability around the starting setting, observing how the network S-parameters are affected. To this regard, the following aspect should always be borne in mind. Given a certain network topology that is kept fixed, there is no unique set of lumped elements' values yielding a good match between the network response and the target datasets in a certain frequency range. Nonetheless, just one among those sets of values sticks better to the physical features of the analysed devices and, therefore, is useful to build behavioural considerations bearing effective significance.

After this articulated discussion, it is worth recalling in brief a statement reported above. Having target devices that are very similar to each other, such as in figure 4.22 (a) and figure 4.22 (b), is particularly favourable to verify the soundness of the equivalent lumped element network topologies. As a matter of fact, the more the architecture is based upon solid physical considerations, the less the network configuration and the lumped elements values need to be modified in order to maintain accuracy. Ideally, still referring to the examples in figure 4.22, the network topology should exclusively require the modification of the contact resistances to reproduce the measured curves of the ON and OFF state, as well as just the reconfiguration of the load resistor, to pass from the series to the shunt attenuator device (figure 4.22 (a) and figure 4.22 (b), respectively).

Extracted networks and discussion of their behaviour

In order to develop equivalent lumped elements' networks, a proper simulation environment is necessary as the first item. As a matter of fact, any tool featuring symbolic composition of analogue circuits, libraries of passive lumped elements based on mathematical models and allowing S-parameters analysis, is suitable to pursue the target discussed here. Just to mention a few commercial software tools, the Advanced Design System (ADS) by Keysight Technologies [78], the Spectre Circuit Simulator within Cadence [79] and the HyperLynx Analog within Mentor [80], are examples of exploitable simulation environments. Some of these tools also offer the possibility to run automated optimization algorithms, thus speeding up the identification of appropriate values for the network lumped elements.

In the examples discussed below, the Quite Universal Circuit Simulation (QUCS) [81] is selected as a software tool for equivalent lumped element networks extraction and analysis. The QUCS is an ICs simulator featuring a Graphical User Interface (GUI) for networks assembly at a schematic level. A fair variety of components' libraries is available, spanning from lumped passives, such as reactive elements and resistances, to waveguide components, again to active semiconductor devices. The QUCS enables various types of simulation analyses, such as large-signal, small-signal and noise behaviour. Finally, an easy-to-use graphical interface allows a fast arrangement of results, according to the needs of the developer. Also, importantly, the QUCS is released under the General Public License (GNU-GPL). Therefore, it can be freely used, shared and modified by the end users.

Having provided such a context, the QUCS equivalent lumped element network schematic corresponding to the series RF-MEMS attenuator module (see figure 4.22 (a)) with the micro-relay in the OFF state (rest position), is shown in figure 4.23. The input/output CPW portions, visible in figure 4.22 (a), are modelled by the CPW symbols, available in the QUCS waveguide components library. The polycrystalline silicon series load resistor is modelled by the R_{lse} component, while direct input/output coupling, taking place as frequency rises, is modelled by the parallel of C_{air} and R_{air}. The lower branch, in parallel to the attenuation load, represents the path across the MEMS suspended membrane. Since the micro-relay is OFF, parasitic capacitance builds across the underlying contact areas and the suspended gold fingers [13]. The latter is modelled by the C_{off} elements, while L_{se} is the series inductance the RF signal encounters when travelling from the input to the output terminations. The network is stimulated with two RF power sources matched to 50 Ω, i.e. P_1 and P_2, respectively. S-parameters analysis is set, and the material properties of the CPW elements, such as substrate resistivity, dielectric constant and thickness of metal, are defined according to the values previously listed in table 4.1.

The QUCS equivalent lumped element network schematic corresponding to the series RF-MEMS attenuator module (see figure 4.22 (a)) with the micro-relay in the ON state (pulled-in), is shown in figure 4.24. In this case, the polycrystalline silicon resistor is shorted by the pulled-in gold MEMS membrane. Therefore, the attenuation is zeroed.

After comparing the network architecture in figure 4.23 (micro-relay OFF) with that in figure 4.24 (micro-relay ON), the only difference is that OFF state parasitic capacitances C_{off} are replaced by ON state series contact resistors R_{on}, as is straightforward to figure out, when thinking about how ohmic RF-MEMS switches behave [82]. The extracted optimal values of all the lumped elements in the networks reported in figure 4.23 and in figure 4.24, as well as those of the networks in figure 4.27 and in figure 4.28, discussed in the following, are listed in table 4.2.

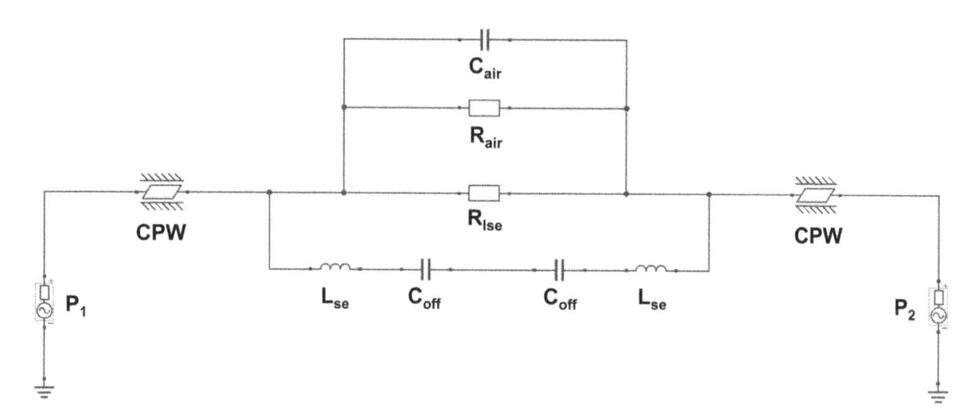

Figure 4.23. Equivalent lumped element network of the series RF-MEMS attenuator module shown in figure 4.22 (a) when the micro-relay is OFF (rest position).

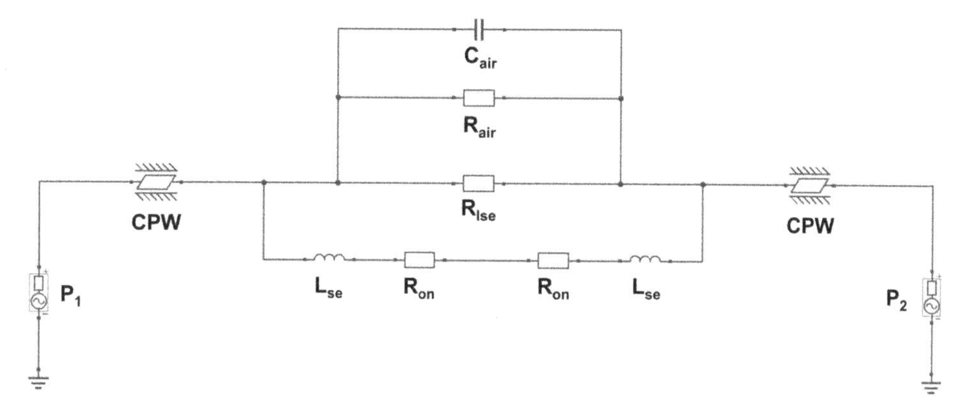

Figure 4.24. Equivalent lumped element network of the series RF-MEMS attenuator module shown in figure 4.22 (a) when the micro-relay is ON (pulled-in).

In the following, the comparison between the target experimental S-parameters traces and the simulated output of extracted lumped elements networks will be discussed. Plots are arranged always according to the same fashion. First, the comparison of the measured and simulated reflection (S11) parameter is reported. Then, the absolute difference (in dB) between the measurement and simulation is plotted for the reflection parameter at both ports (S11 and S22). Furthermore, the same comparisons are reported for the transmission parameter. To this purpose, the measured and simulated loss/attenuation (S21) is reported, together with the absolute difference of measurements and simulations related to both transmission parameters (S12 and S21). In light of the explanation just developed, the results related to the series RF-MEMS attenuator module in the OFF state—device in figure 4.22 (a); lumped network in figure 4.23—are plotted in figure 4.25.

From a qualitative point of view, the equivalent lumped element network reproduces quite satisfactorily both the measured reflection (S11) and attenuation (S21) characteristics, as shown in figure 4.25 (a) and figure 4.25(c), respectively. The

Table 4.2. Summary of the optimal values extracted for all the lumped elements included in the equivalent networks reported in figure 4.23, figure 4.24, figure 4.27 and figure 4.28.

Element name	Description	Value
C_{air}	Direct input/output capacitive coupling through air	12 fF
R_{air}	Direct input/output resistive coupling (loss) through air	10 GΩ
L_{se}	Series inductance of the RF signal line	70 pH
R_{on}	Metal-to-metal resistance between the pulled-in MEMS and the contact area	100 mΩ
C_{off}	Capacitive coupling between the MEMS (rest position) and the contact area	7 fF
R_{lse}	Polycrystalline silicon series resistor	185 Ω
R_{up}	Series resistance of the RF signal underpass	500 mΩ
L_{up}	Series inductance of the RF signal underpass	150 pH
L_{mems}	Inductance of the MEMS micro-switch	40 pH
R_{lsh}	Polycrystalline silicon shunt resistor	42 Ω
W_{CPW}	Width of the input/output CPW signal line	56 μm
G_{CPW}	Gap of the input/output CPW signal line	50 μm
L_{CPW}	Length of the input/output CPW signal line	200 μm

upward step, taking place in the measured reflection between 50 GHz and 60 GHz, corresponds to a downward step in the attenuation curve. Such a characteristic is not present in the simulated curves. However, it must be observed that the absolute error of simulations versus measurements is never too large. Looking at the plot in figure 4.25 (b), the difference between the experimental and simulated reflection (both concerning S11 and S22) is below 1 dB in most of the frequency range, while it is below 3 dB in two sub-ranges centred around 50 GHz and 100 GHz. However, the difference of attenuation (both concerning S12 and S21) in figure 4.25 (d) is below 1 dB up to 55 GHz, below 2 dB up to 100 GHz and below 3 dB up to 110 GHz.

The results related to the series RF-MEMS attenuator module in the ON state—device in figure 4.22 (a); lumped network in figure 4.24—are plotted in figure 4.26.

Also in this circumstance, the qualitative prediction operated by the equivalent lumped element network is quite satisfactory, for both the reflection (S11) and loss (S21) characteristics, reported in figure 4.26 (a) and in figure 4.26 (c), respectively. In particular, the error of the simulated loss (S12 and S21) is quite low, especially up to around 100 GHz, as shown in figure 4.26 (d). In contrast, the difference between the simulated and measured reflection (S11 and S22) is more pronounced, as visible in figure 4.26 (b). Nonetheless, when the RF-MEMS device is conducting, the reflection scores larger values, in the order of a few tens of dB. Therefore, larger error levels can be expected and tolerated. Moreover, since equivalent networks enable behavioural approximate description of physical devices, it is quite difficult to obtain an accurate prediction of all the S-parameters, especially on a very wide frequency range, such as the one considered here, spanning from 200 MHz up to

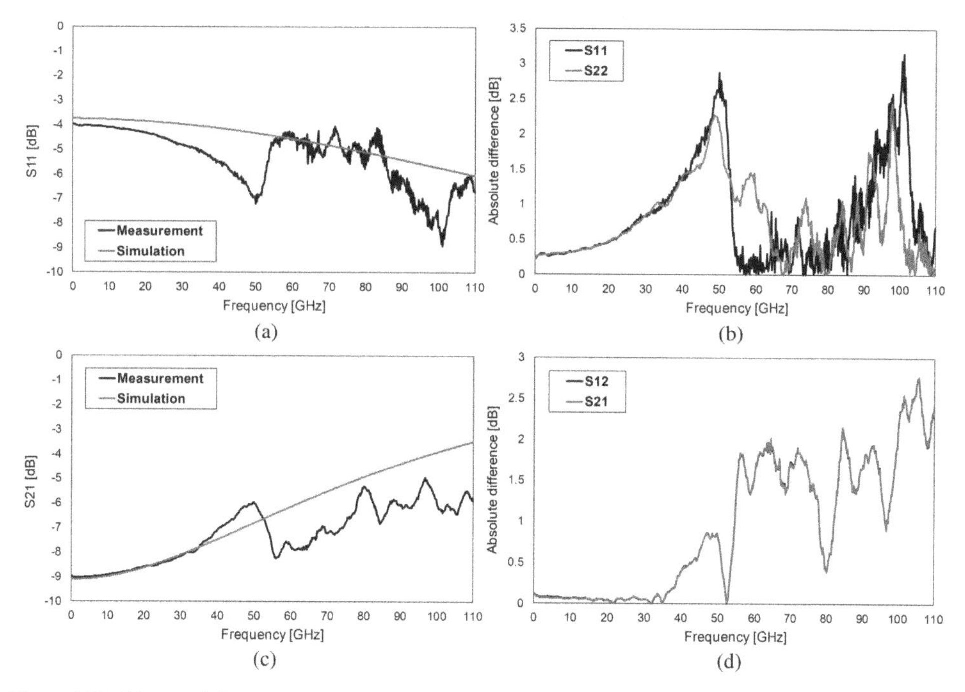

Figure 4.25. Measured S-parameters characteristics of the series RF-MEMS attenuator module in figure 4.22 (a) when the micro-relay is OFF and response of the equivalent circuit in figure 4.23, from 200 MHz up to 110 GHz. (a) Measured versus simulated reflection (S11) characteristic. (b) Absolute difference (in dB) between the measured and simulated reflection characteristics at both ports (S11 and S22). (c) Measured versus simulated attenuation (S21) characteristic. (d) Absolute difference (in dB) between the measured and simulated attenuation characteristics at both ports (S12 and S21).

110 GHz. Still referring to figure 4.26, it is possible to tailor the lumped elements' values in order to achieve a better match of reflection. However, this would bring a larger error of the transmission parameters. In this case, the accuracy of S12 and S21 parameters was considered to be more critical than that of S11 and S22.

Stepping forth, the QUCS equivalent lumped element network schematic corresponding to the shunt RF-MEMS attenuator module (see figure 4.22 (b)) with the micro-relay in the OFF state (rest position), is shown in figure 4.27. In this configuration, the resistive load is not selected and the device does not realize any RF signal attenuation.

The comparison between the equivalent networks of the series and shunt RF-MEMS attenuation modules, reported in figure 4.23 and in figure 4.27, respectively, highlights their similarities and differences. On a general reference plane, the architectures are very similar. The series attenuation resistive load is replaced in figure 4.27 by the resistive (R_{up}) and inductive (L_{up}) series contributions of the underpass connecting the input/output terminations, visible in figure 4.22 (b). Moreover, a branch to RF ground accounts for the shunt attenuating resistor. It features the inductance of the path to ground (L_{mems}) and, of course, the polycrystalline silicon shunt resistive load (R_{lsh}). In addition, the QUCS equivalent lumped

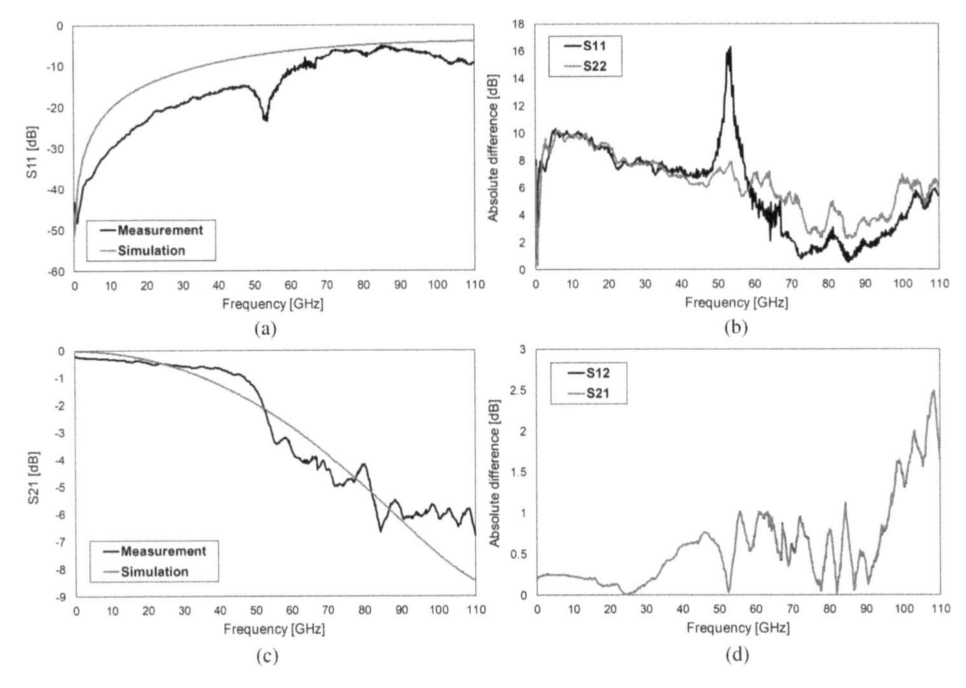

Figure 4.26. Measured S-parameters characteristics of the series RF-MEMS attenuator module in figure 4.22 (a) when the micro-relay is ON and response of the equivalent circuit in figure 4.24, from 200 MHz up to 110 GHz. (a) Measured versus simulated reflection (S11) characteristic. (b) Absolute difference (in dB) between the measured and simulated reflection characteristics at both ports (S11 and S22). (c) Measured versus simulated loss (S21) characteristic. (d) Absolute difference (in dB) between the measured and simulated loss characteristics at both ports (S12 and S21).

element network schematic corresponding to the shunt RF-MEMS attenuator module (see figure 4.22 (b)) with the micro-relay in the ON state (pulled-in), is shown in figure 4.28.

For the case of the series RF-MEMS device, when referring to the shunt device the micro-relay OFF and ON states are modelled simply by replacing parasitic series capacitances (C_{off} in figure 4.27) with series contact resistances (R_{on} in figure 4.28). Of course, when C_{off} capacitors are inserted, the shunt branch featuring the resistive load does not influence the S-parameters' behaviour of the network. However, when the R_{on} resistances are included in the architecture, part of the RF signal is shorted to ground through R_{lsh}. The results related to the shunt RF-MEMS attenuator module in the OFF state—device in figure 4.22 (b); lumped network in figure 4.27—are plotted in figure 4.29.

The match between the measured and simulated reflection (S11) characteristic is very good up to around 60 GHz, but gets worse above that frequency, as shown in the S11 plot in figure 4.29 (a) as well as in the error plot of (S11 and S22) reported in figure 4.29 (b). In detail, with reference to the latter graph, the difference is below 2 dB up to 60 GHz, while in the upper part of the frequency range it increases up to around 16 dB. When looking at the loss (S21) characteristic, a satisfactory match

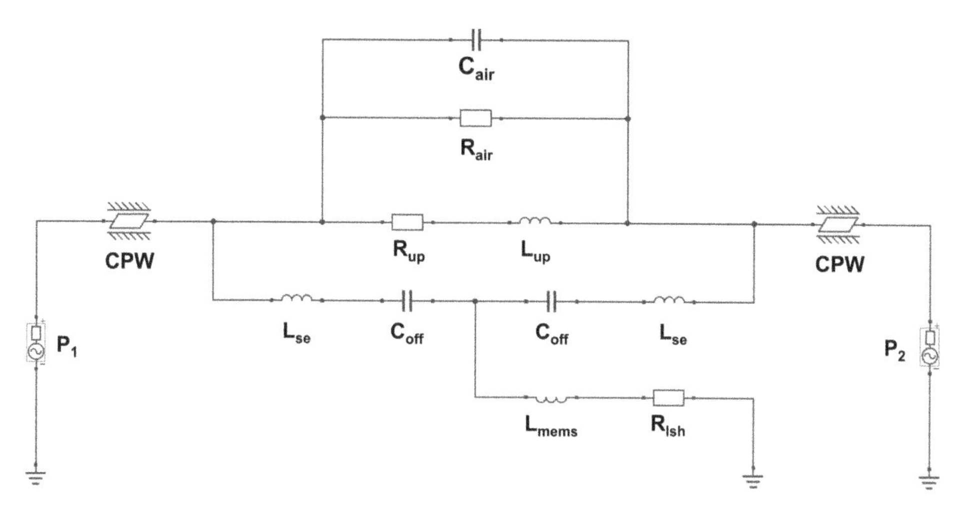

Figure 4.27. Equivalent lumped element network of the shunt RF-MEMS attenuator module shown in figure 4.22 (b) when the micro-relay is OFF (rest position).

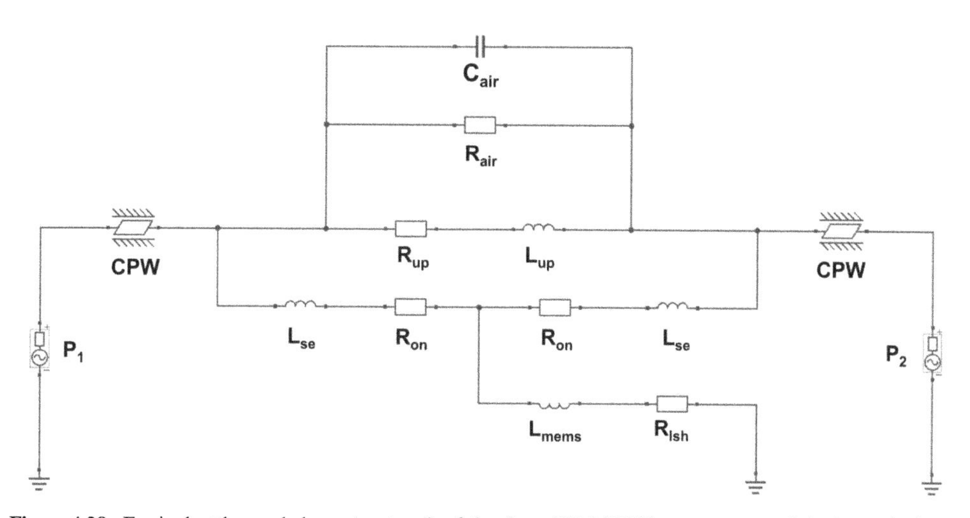

Figure 4.28. Equivalent lumped element network of the shunt RF-MEMS attenuator module shown in figure 4.22 (b) when the micro-relay is ON (pulled-in).

between simulations and measurements is noticeable in figure 4.29 (c). The differences between the transmission parameters (S12 and S21) are quite limited up to 80 GHz, while from 80 GHz to 110 GHz increase to the worst value of about 8 dB.

Finally, the results related to the shunt RF-MEMS attenuator module in the ON state—device in figure 4.22 (b); lumped network in figure 4.28—are plotted in figure 4.30. In this circumstance, from a general point of view, the match between the measurements and simulations is not as good as in the cases discussed previously.

The qualitative match between the measured and simulated reflection (S11) characteristics, reported in figure 4.30 (a), is not fully satisfactory. While the

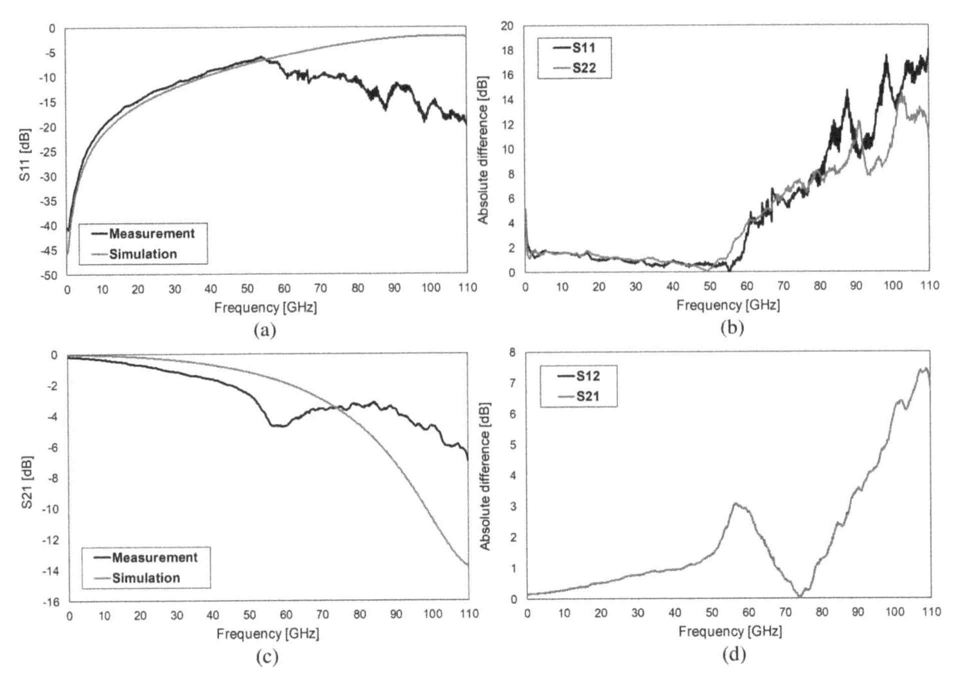

Figure 4.29. Measured S-parameters' characteristics of the shunt RF-MEMS attenuator module in figure 4.22 (b) when the micro-relay is OFF and response of the equivalent circuit in figure 4.27, from 200 MHz up to 110 GHz. (a) Measured versus simulated reflection (S11) characteristic. (b) Absolute difference (in dB) between the measured and simulated reflection characteristics at both ports (S11 and S22). (c) Measured versus simulated loss (S21) characteristic. (d) Absolute difference (in dB) between the measured and simulated loss characteristics at both ports (S12 and S21).

measurement trace is substantially flat, the simulated one bends upward as the frequency increases. On the other hand, looking at the attenuation (S21) characteristic in figure 4.30 (c), a good match is visible only up to 60 GHz. Observing the plots related to the difference between the measurements and simulations, both the one concerning reflection (S11 and S22) and attenuation (S12 and S21) characteristics—figure 4.30 (b) and figure 4.30 (d), respectively—exhibit reasonably limited value up to around 60 GHz only.

In conclusion, the methodology just discussed for the definition and extraction of equivalent lumped element networks of RF-MEMS devices provides a significant helping hand to the developer/designer. As reported in the previous pages, it is possible to extract network architectures holding validity on very wide frequency ranges. To this scope, the observed range from 200 MHz up to 110 GHz is extremely wide, and it was employed mainly for explanatory purposes, with the aim of challenging quite harshly the robustness of the discussed approach. In real cases, what the RF-MEMS developer needs are lumped networks exhibiting good accuracy in a range from 5–10 GHz up to 20–30 GHz at most. In light of the plots previously commented on and reported in figure 4.25, figure 4.26, figure 4.29 and figure 4.30, it clearly emerges that restricting the analysed frequency span from 110

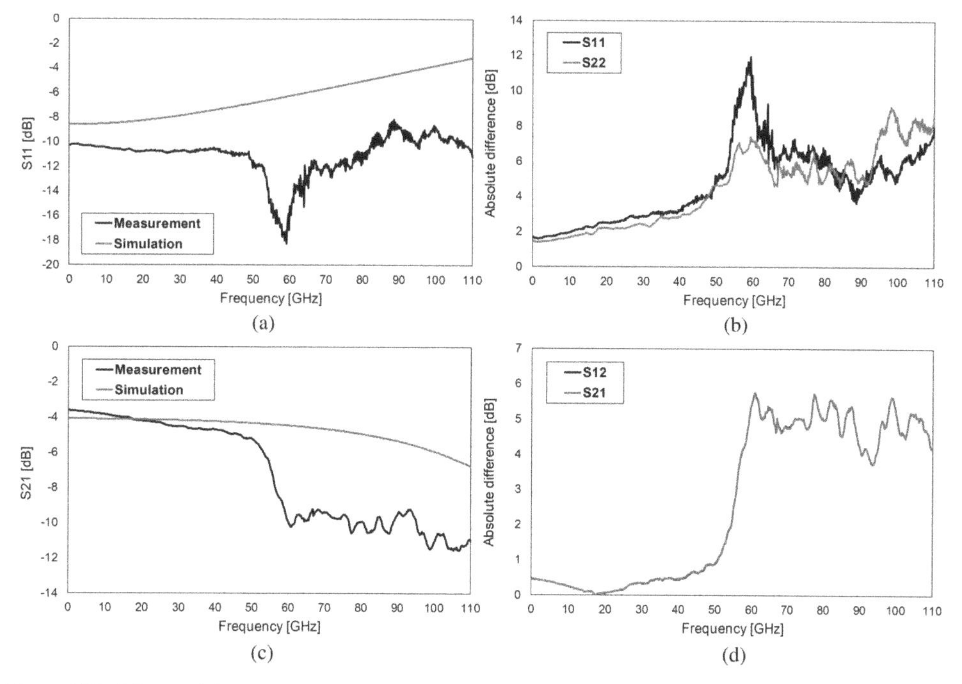

Figure 4.30. Measured S-parameters' characteristics of the shunt RF-MEMS attenuator module in figure 4.22 (b) when the micro-relay is ON and response of the equivalent circuit in figure 4.28, from 200 MHz up to 110 GHz. (a) Measured versus simulated reflection (S11) characteristic. (b) Absolute difference (in dB) between the measured and simulated reflection characteristics at both ports (S11 and S22). (c) Measured versus simulated attenuation (S21) characteristic. (d) Absolute difference (in dB) between the measured and simulated attenuation characteristics at both ports (S12 and S21).

GHz down to a few tens of GHz would lead to significantly improved levels of accuracy.

Conclusion

This conclusive chapter builds on the discussion previously developed on future 5G applications, presenting the analysis of RF-MEMS passive components by means of more practical and hands-on approaches. Nonetheless, before stepping into detail, a comprehensive analysis of target performances was developed. As a matter of fact, the preceding chapter 3 on the 5G scenario, listed system-level characteristics that mobile networks and infrastructures will have to satisfy. Therefore, the initial part of this chapter was dedicated to deriving a plausible set of specifications, at device level, that RF passive components should be able to reach, in order to be suitable for 5G applications. Providing these numbers, case studies and examples of RF-MEMS design optimization were reviewed.

First, the focus was concentrated on the coupled electromechanical characteristics of MEMS devices. To this regard, a few switching unit topologies were placed under the spotlight, capitalizing on Finite Element Method (FEM) simulations aimed at investigating their characteristics and performance. In particular, the influence of the flexible suspensions geometry on the pull-in and pull-out voltages was analysed in static conditions. Therefore, the dynamic behaviour of RF-MEMS switches was also studied, showing how the mechanical stiffness of suspensions can influence the opening/closing switching time in transient conditions. In particular, it was highlighted how the RF-MEMS designer very often has to deal with conflicting specifications, and how proper trade-offs must be sought, trying to achieve satisfactory characteristics without impairment of other device features, at the same time. In particular, it was demonstrated that lowering the pull-in voltage leads to MEMS switches with slower actuated/release dynamic response. To complete the discussion related to electromechanical modelling of RF-MEMS, the analysis approach based on compact analytical models was also reported. Methodologies were discussed on the advantages and disadvantages of FEM versus compact models, focusing on the specific circumstances in which each of them performs better and, therefore, is a more appropriate option.

Then, discussion around the electromagnetic modelling of RF-MEMS was developed. At first, the selected FEM tool underwent a comprehensive validation phase, taking experimental S-parameters' datasets as reference targets. The validation was necessary to ensure accurate prediction of the physical device's behaviour, through tailoring the 3D model assembly and the simulator settings on the actual characteristics of a real selected technology platform. Furthermore, the validated electromagnetic simulation method was employed to discuss a few RF-MEMS switching unit design concepts. In particular, the influence of the suspended membrane and of the contacts geometry on critical parameters like isolation (when OPEN) and loss (when CLOSE) was discussed. As previously done concerning the electromechanical behaviour, the presence of conflicting characteristics in the electromagnetic domain was reported. In particular, the trade-off between isolation and loss was analysed, also introducing the increase of redundancy as a possible strategy to improve the RF-MEMS device characteristics as a whole.

Finally, the discussion on RF modelling was completed by an extensive overview of a methodology for extracting RF-MEMS equivalent lumped element networks. Such architectures are very useful both to comprehend in-depth the physical causes lying behind the RF behaviour of the micro-devices, as well as to easily parameterize and/or co-simulate the MEMS part together with the rest of the system/sub-system. Eventually, all the reported characteristics and performance of the selected RF-MEMS case studies were commented on against the set of 5G target specifications mentioned in the beginning of the chapter, thus closing the whole discussion developed throughout this work.

References

[1] Giacomozzi F, Mulloni V, Colpo S, Iannacci J, Margesin B and Faes A 2011 A flexible fabrication process for RF MEMS devices *Rom. J. Inf. Sci. Technol* **14** 259–68

[2] Iannacci J 2013 *Practical Guide to RF-MEMS* 1st edn (Weinheim: Wiley) p 372

[3] Iannacci J and Gaddi R 2010 *Mixed-Domain Simulation and Wafer-Level Packaging of RF-MEMS Devices* 1st edn (Saarbrücken: Lambert Academic) p 282

[4] Senturia S D 2001 *Microsystem Design* 1st edn (New York: Springer) p 689
[5] Mentor *Tanner* L-Edit IC *Layout* www.mentor.com/tannereda/l-edit [accessed 17 July 2017]
[6] Culpepper M L *Creating DXF Files For The Waterjet* https://ocw.mit.edu/courses/mechanical-engineering/2-000-how-and-why-machines-work-spring-2002/tools/2_000_DXF_Tutorial.pdf [accessed 17 July 2017]
[7] *FreeCAD* www.freecadweb.org/ [accessed 17 July 2017]
[8] *The FreeCAD Manual* www.freecadweb.org/wiki/Import_Export [accessed 17 July 2017]
[9] ANSYS *Workbench Platform* www.ansys.com/products/platform [accessed 17 July 2017]
[10] Harrar D II *MEMS Multiphysics Simulation in ANSYS Workbench* www.ozeninc.com/downloads/PRESENTATION-Mulitiphysics_Simulation_for_MEMS_Using_Workbench.pdf [accessed 17 July 2017]
[11] Iannacci J, Gaddi R and Gnudi A 2007 Non-linear electromechanical RF model of a MEMS varactor based on VerilogA© and lumped-element parasitic network *Proc. 37th Europ. Microw. Conf. EuMC (Munich, Oct. 2007)* pp 544–7
[12] Pacheco S P, Katehi L P B and Nguyen C T C 2000 Design of low actuation voltage RF MEMS switch *Proc. IEEE MTT-S Int. Microw. Symp. (Boston, MA, June 2000)* **1** 165–8
[13] Iannacci J, Gaddi R and Gnudi A 2010 Experimental validation of mixed electromechanical and electromagnetic modeling of RF-MEMS devices within a standard IC simulation environment *IEEE J. Microelectromechan. Syst.* **19** 526–37
[14] *ANSYS—Cantilever Beam Modal Analysis* https://confluence.cornell.edu/display/SIMULATION/ANSYS+-+Cantilever+Beam+Modal+Analysis [accessed 3 August 2017]
[15] Sun Y, Fang D and Soh A K 2006 Thermoelastic damping in micro-beam resonators *Int. J. Solids Struct.* **43** 3213–29
[16] Künzig T, Niessner M, Wachutka G, Schrag G and Hammer H 2010 The effect of thermoelastic damping on the total Q-factor of state-of-the-art MEMS gyroscopes with complex beam-like suspensions *Procedia Eng.* **5** 1296–9
[17] De S K and Aluru N R 2006 Theory of thermoelastic damping in electrostatically actuated microstructures *Phys. Rev.* B **74** 1–13
[18] Iannacci J and Gaddi R 2010 *Mixed-Domain Simulation and Wafer-Level Packaging of RF-MEMS Devices* 1st edn (Saarbrücken: Lambert Academic) p 282
[19] Iannacci J, Del Tin L, Gaddi R, Gnudi A and Rangra K J 2005 Compact modeling of a MEMS toggle-switch based on modified nodal analysis *Proc. Symp. on Design, Test, Integration and Packaging of MEMS/MOEMS DTIP 2005 (Montreux, June 2005)* pp 411–6
[20] Iannacci J and Gaddi R 2005 Nodal modelling of uneven electrostatic transduction in MEMS ed C Di Natale, A D'Amico, G Martinelli, M C Carotta and V Guidi *Proc. 9th Italian Conf. on Sensors and Microsystems* (Singapore: World Scientific) pp 385–9
[21] Niessner M, Schrag G, Iannacci J and Wachutka G 2011 Macromodel-based simulation and measurement of the dynamic pull-in of viscously damped RF-MEMS switches *Sensors Actuators* A **172** 269–79
[22] Fruehling A, Yang W and Peroulis D 2012 Cyclic evolution of bouncing for contacts in commercial RF MEMS switches *Proc. IEEE 25th Int. Conf. on Micro Electro Mechanical Syst. MEMS (Paris, Jan. 2012)* pp 688–91
[23] Peschot A, Poulain C, Valadares C, Reig B, Bonifaci N and Lesaint O 2014 Evolution of contact bounces in MEMS switches under cycling *Proc. IEEE 60th Holm Conf. on Electrical Contacts (New Orleans, LA, Oct. 2014)* pp 1–6

[24] Tavassolian N, Koutsoureli M, Papaioannou G, Lacroix B and Papapolymerou J 2010 Dielectric charging in capacitive RF MEMS switches: the effect of electric stress *Proc. Asia-Pacific Microw. Conf. APMC (Yokohama, Dec. 2010)* pp 1833–6

[25] Zaghloul U, Bhushan B, Pons P, Papaioannou G, Coccetti F and Plana R 2011 Different stiction mechanisms in electrostatic MEMS devices: nanoscale characterization based on adhesion and friction measurements *Proc. Int. Solid-State Sens., Actuators and Microsystems Conf. TRANSDUCERS (Beijing, June 2011)* pp 2478–81

[26] Tazzoli A, Iannacci J and Meneghesso G 2011 A positive exploitation of ESD events: micro-welding induction on ohmic MEMS contacts *Proc. Electrical Overstress/Electrostatic Discharge Symp. EOS/ESD (Anaheim, CA, Sept. 2011)* pp 1–8

[27] Iannacci J, Faes A, Repchankova A, Tazzoli A and Meneghesso G 2011 An active heat-based restoring mechanism for improving the reliability of RF-MEMS switches *Microelectron. Reliab.* **51** 1869–73

[28] *Verilog-A Language Reference Manual. Analog Extensions to Verilog HDL* www.siue.edu/~gengel/ece585WebStuff/OVI_VerilogA.pdf [accessed 23 July 2017]

[29] *Cadence* www.cadence.com/ [accessed 24 July 2017]

[30] Iannacci J, Faes A, Kuenzig T, Niessner M and Wachutka G 2011 Electromechanical and electromagnetic simulation of RF-MEMS complex networks based on compact modeling approach *Proc. TechConnect World, NSTI Nanotech (Boston, MA, June 2011)* pp 591–4

[31] Iannacci J, Faes A, Mastri F, Masotti D and Rizzoli V 2010 A MEMS-based wide-band multi-state power attenuator for radio frequency and microwave applications *Proc. TechConnect World, NSTI Nanotech (Anaheim, CA, June 2010)* pp 328–31

[32] *Spectre Circuit Simulator* www.cadence.com/content/cadence-www/global/en_US/home/tools/custom-ic-analog-rf-design/circuit-simulation/spectre-circuit-simulator.html [accessed 24 July 2017]

[33] Conway J A, Osborn J V and Fowler J D 2007 Stroboscopic imaging interferometer for MEMS performance measurement *IEEE J. Microelectromech. Syst.* **16** 668–74

[34] Hart M R, Conant R A, Lau K Y and Muller R S 2000 Stroboscopic interferometer system for dynamic MEMS characterization *IEEE J. Microelectromech. Syst.* **9** 409–18

[35] Fedder G K and Jing Q 1999 A hierarchical circuit-level design methodology for micro-electromechanical systems *IEEE Trans. Circ. Syst. II: Analog and Digital Signal Process.* **46** 1309–15

[36] Jing Q, Mukherjee T and Fedder G K 2002 Schematic-based lumped parameterized behavioral modeling for suspended MEMS *Proc. IEEE/ACM Int. Conf. on Computer Aided Design, 2002 ICCAD (San Jose, CA, Nov. 2002)* pp 367–73

[37] Ciuprina G, Ioan D, Lup A S, Popescu M, Barbulescu R and Stefanescu A 2016 Coupled multiphysics-RF reduced models for MEMS *Proc. IEEE 1st Int. Conf. on Power Electron., Intelligent Control and Energy Syst. ICPEICES (Delhi, July 2016)* pp 1–6

[38] Halder S, Palego C, Peng Z, Hwang J C M, Forehand D I and Goldsmith C L 2009 Compact RF model for transient characteristics of MEMS capacitive switches *IEEE Trans. Microw. Theory Tech.* **57** 237–42

[39] Veijola T, Tinttunen T, Nieminen H, Ermolov V and Ryhanen T 2002 Gas damping model for a RF MEM switch and its dynamic characteristics *Proc. IEEE MTT-S Int. Microw. Symp. (Seattle, WA, June 2002)* pp 1213–6

[40] Veijola T and Lehtovuori A 2008 Numerical and compact modelling of squeeze-film damping in RF MEMS resonators *Proc. DTIP Symp. on Design, Test, Integration and Packaging of MEMS/MOEMS (Nice, April 2008)* pp 222–8
[41] Zhou N, Clark J V and Pister K S J 1998 Nodal simulation for MEMS design using SUGAR v0.5 *Proc. Int. Conf. on Modeling and Simulation of Microsystems Semiconductors, Sensors and Actuators (Santa Clara, CA, April 1998)* pp 308–13
[42] Rappaport T S, Sun S, Mayzus R, Zhao H, Azar Y, Wang K, Wong G N, Schulz J K, Samimi M and Gutierrez F 2002 Addressing the needs of complex MEMS design *Proc. IEEE Int. Conf. on Micro Electro Mechanical Systems MEMS (Las Vegas, NV, Jan. 2002)* pp 204–9
[43] *Nodal Design of Actuators and Sensors: Nodasv1.4* www.ece.cmu.edu/~mems/projects/memsyn/nodasv1_4/ [accessed 24 July 2017]
[44] Wong G C, Tse G K, Jing Q, Mukherjee T and Fedder G K 2003 Accuracy and composability in NODAS *Proc. IEEE Int. Workshop on Behavioral Modeling and Simulation BMAS (San Jose, CA, Oct. 2003)* pp 82–7
[45] Jing Q, Mukherjee T and Fedder G K 2002 Schematic-based lumped parameterized behavioural modeling for suspended MEMS *Proc. IEEE/ACM Int. Conf. on Computer Aided Design ICCAD (San Jose, CA, Nov. 2002)* pp 367–73
[46] Guo C and Fedder G K 2011 2-DoF twisting electrothermal actuator for scanning laser rangefinder application *Proc. IEEE Int. Conf. on Micro Electro Mechanical Syst. (Cancun, Jan. 2011)* pp 1205–8
[47] Jing Q, Luo H, Mukherjee T, Carley L R and Fedder G K 2000 CMOS micromechanical bandpass filter design using a hierarchical MEMS circuit library *Proc. IEEE Annu. Int. Conf. on Micro Electro Mechanical Systems (Miyazaki, Jan. 2000)* pp 187–92
[48] Tatar E, Mukherjee T and Fedder G K 2014 Simulation of stress effects on mode-matched MEMS gyroscope bias and scale factor *Proc. IEEE/ION Position, Location and Navigation Symp. PLANS (Monterey, CA, May 2014)* pp 16–20
[49] Schrag G, Brand O, Fedder G K, Hierold C and Korvink J G (ed) 2013 *System-level Modeling of MEMS* 1st edn (Weinheim: Wiley) p 562
[50] Ansys HFSS www.ansys.com/products/electronics/ansys-hfss [accessed 25 July 2017]
[51] Casini F, Farinelli P, Mannocchi G, DiNardo S, Margesin B, De Angelis G, Marcelli R, Vendier O and Vietzorreck L 2010 High performance RF-MEMS SP4T switches in CPW technology for space applications *Proc. European Microw. Conf. EuMW (Paris, Sep.–Oct. 2010)* pp 89–92
[52] Diaferia F, Deborgies F, Di Nardo S, Espana B, Farinelli P, Lucibello A, Marcelli R, Margesin B, Giacomozzi F, Vietzorreck L and Vitulli F 2014 Compact 12×12 switch matrix integrating RF MEMS switches in LTCC hermetic packages *Proc. European Microw. Conf. EuMW (Rome, Oct. 2014)* pp 199–202
[53] Giacomozzi F, Mulloni V, Colpo S, Faes A, Sordo G and Girardi S 2013 RF-MEMS devices packaging by using quartz caps and epoxy polymer sealing rings *Proc. Symp. on Design, Test, Integration and Packaging of MEMS/MOEMS DTIP (Barcelona, April 2013)* pp 1–6
[54] Iannacci J 2013 Simulation techniques (commercial tools) *Practical Guide to RF-MEMS* 1st edn (Weinheim: Wiley) ch 4 pp 121–31
[55] Iannacci J and Tschoban C 2017 RF-MEMS for future mobile applications: experimental verification of a reconfigurable 8-bit power attenuator up to 110 GHz *J. Micromech. Microeng.* **27** 1361–6439

[56] Huang J T, Hsu Y K, Lo Y C, Lee K Y, Chen C K and Tsai T C 2010 Design and fabrication of low-insertion loss and high-isolation CMOS-MEMS switch for microwave applications *Proc. Int. Microsystems Packaging Assembly and Circuits Technol. Conf. (Taipei, Oct. 2010)* pp 1–3

[57] Fomani A A and Mansour R R 2009 Miniature RF MEMS switch matrices *Proc. IEEE MTT-S Int. Microw. Symp. (Boston, MA, June 2009)* pp 1221–4

[58] Shen H, Gong S and Barker N S 2008 DC-contact RF MEMS switches using thin-film cantilevers *Proc. Europ. Microw. Integrated Circuit Conf. EuMIC (Amsterdam, Oct. 2008)* pp 382–5

[59] Kanthamani S, Raju S and Abhai Kumar V 2008 Design of low actuation voltage RF MEMS cantilever switch *Proc. Int. Conf. on Recent Adv. in Microw. Theory and Applications (Jaipur, Nov. 2008)* pp 584–6

[60] Muley C A and Naveed S A 2013 Modelling of cantilever based MEMS RF switch *Proc. Int. Conf. on Computing, Commun. and Networking Technol. ICCCNT (Tiruchengode, July 2013)* pp 1–5

[61] Shalaby M M, Wang Z, Chow L L-W, Jensen B D, Volakis J L, Kurabayashi K and Saitou K 2009 Robust design of RF-MEMS cantilever switches using contact physics modeling *IEEE Trans. Ind. Electron.* **56** 1012–21

[62] Homayouni S M, Schreurs D and Nauwelaers B 2009 Wide-band multi-bias equivalent circuit extraction for FinFET transistors accommodating high-frequency kink-behaviours *Proc. Europ. Microw. Integrated Circuits Conf. EuMIC (Rome, Sept. 2009)* pp 85–8

[63] Wada T, Nakajima R, Obiya H, Ogami T, Koshino M, Kawashima M and Nakajima N 2014 A miniaturized broadband lumped element circulator for reconfigurable front-end system *Proc. IEEE MTT-S Int. Microw. Symp. IMS2014 (Tampa, FL, June 2014)* pp 1–3

[64] Essaadali R, Jarndal A, Kouki A B and Ghannouchi F M 2016 A new GaN HEMT equivalent circuit modeling technique based on X-parameters *IEEE Trans. Microw. Theory Tech.* **64** 2758–77

[65] Arcioni P, Castello R, Perregrini L, Sacchi E and Svelto F 1998 An improved lumped-element equivalent circuit for on silicon integrated inductors *Proc. IEEE Radio and Wireless Conf. RAWCON (Colorado Springs, CO, Aug. 1998)* pp 301–4

[66] Kang J, Sun L, Wen J and Zhao M 2009 An equivalent lumped-circuit model for on-chip symmetric intertwined transformer *Proc. IEEE Int. Conf. on ASIC (Changsha, Hunan, Oct. 2009)* pp 674–7

[67] Horng T S, Jau J K, Tsai Y S and Huang C S A decomposition and reconstruction scheme for broadband modeling of on-chip passive components using the modified T-equivalent circuit topology *Proc. IEEE Radio Frequency integrated Circuits (RFIC) Symp (Long Beach, CA, June 2005)* pp 299–302

[68] Hwang J, Jung W and Kim S 2015 Coupling analysis and equivalent circuit model of the IC stripline method *Proc. Asia-Pacific Symp. on Electromag. Compatibility APEMC (Taipei, May 2015)* pp 650–3

[69] Zhou Y and Chen Y 2008 Lumped-element equivalent circuit models for distributed microwave directional couplers *Proc. Int. Conf. on Microw. and Millimeter Wave Technol. (Nanjing, April 2008)* pp 131–4

[70] Yang K, Yin W Y and Mao J F 2007 Wideband lumped element model for on-chip (a) symmetrical coupled interconnects on lossy silicon substrate *Proc. Asia-Pacific Microw. Conf. (Bangkok, Dec. 2007)* pp 1–4

[71] Chen K L, Salvia J, Potter R, Howe R T and Kenny T W 2009 Performance evaluation and equivalent model of silicon interconnects for fully-encapsulated RF MEMS devices *IEEE Trans. Adv. Packag.* **32** 402–9
[72] Grinde C, Nygaard K H, Due-Hansen J and Fjeldly T A 2009 Lumped modeling of a novel RF MEMS double-disk resonator system *Proc. Symp. on Design, Test, Integration & Packag. of MEMS/MOEMS DTIP (Rome, April 2009)* pp 14–8
[73] Unlu M, Topalli K, Demir S, Civi O A, Koc S and Akin T 2004 A parametric modeling study on distributed MEMS transmission lines *Proc. Europ. Microw. Conf. EuMW (Amsterdam, Oct. 2004)* pp 1157–60
[74] Klymyshyn D M, Martin Borner M, Haluzan D T, Santosa E G, Melissa Schaffer M, Achenbach S and Mohr J 2010 Vertical hgh-Q RF-MEMS devices for reactive lumped-element circuits *IEEE Trans. Microw. Theory Techn.* **58** 2976–86
[75] Iannacci J 2014 RF-MEMS: A development flow driving innovative device concepts to high performance components and networks for wireless applications *Proc. Microelectron. Syst. Symp. MESS (Vienna, May 2014)* pp 1–6
[76] Iannacci J, Huhn M, Tschoban C and Pötter H 2016 RF-MEMS technology for 5G: series and shunt attenuator modules demonstrated up to 110 GHz *IEEE Electron Device Lett.* **37** 1336–9
[77] Iannacci J, Reyes J, Maaß U, Huhn M, Ndip I and Pötter H 2016 RF-MEMS for 5G mobile communications: A basic attenuator module demonstrated up to 50 GHz *Proc. IEEE SENSORS (Orlando, FL, Oct.–Nov. 2016)* pp 1–3
[78] Keysight Technologies *Advanced Design System (ADS)* www.keysight.com/en/pc-1297113/advanced-design-system-ads?c.c.=IT&lc=ita [accessed 1 August 2017
[79] Cadence *Spectre Circuit Simulator* www.cadence.com/content/cadence-www/global/en_US/home/tools/custom-ic-analog-rf-design/circuit-simulation/spectre-circuit-simulator.html [accessed 1 August 2017]
[80] Mentor *HyperLynx Analog* www.mentor.com/pcb/hyperlynx/analog/ [accessed 1 August 2017]
[81] *Quite universal Circuit Simulator* http://qucs.sourceforge.net/ [accessed 2 August 2017]
[82] Iannacci J 2010 Mixed-domain fast simulation of RF and microwave MEMS-based complex networks within standard IC development frameworks *Advanced Microwave Circuits and Systems* ed V Zhurbenko 1st edn (Rijeka: INTECH) pp 313–38

Appendix A

Moving MEMS: Dynamics down in the micro-world

A.1 Introduction

The aim of this appendix is to provide a brief insight into MEMS and RF-MEMS dynamics, by reporting and explaining a few video animations of Microsystem devices caught during their operation. The material will be based both on experimental measurements as well as Finite Element Method (FEM) simulations. The way sections are split is based upon the type of measurement setup or simulation exploited to obtain visual dynamic data. According to this arrangement, the necessary background on the observation procedure is provided first, in such a way as to ease understanding.

A.2 Measurements based on interferometric microscopy

The experimental material that is going to be reported in this section is acquired by means of a setup based on optical interferometry. This particular type of microscopy analysis exploits the constructive/destructive interference of the light coming back from the Device Under Test (DUT) with that reflected by a mirror, in order to reconstruct information about the vertical dimension of the DUT itself [1–2]. In other words, optical interferometry allows one to reconstruct the 3D profile of the DUT, starting from single microscopy observation. Proper software-based post-processing of the measured data enables the 3D images rendering of the DUT in which a colour scale is typically used to provide information about the vertical quote of each measured spot. The image reported in figure A.1 shows the 3D topology of an RF-MEMS two-state capacitor obtained with an optical interferometer based on white light. The DUT is a typical MEMS suspended electrostatically controlled plate, framed within a Coplanar Waveguide (CPW) structure.

doi:10.1088/978-0-7503-1545-6ch5

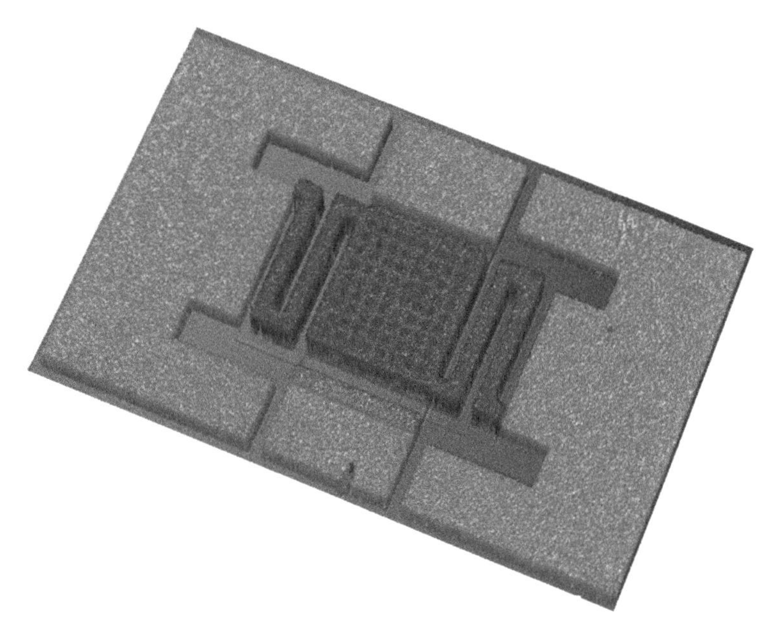

Figure A.1. 3D image of an RF-MEMS two-state capacitor obtained by means of optical interferometry. The colour scale indicates the vertical quote of the DUT.

Looking at the colour scale of the image in figure A.1, three main reference quotes can be easily identified. The lower plane, depicted in blue, is the underlying substrate. The intermediate quote, reported in green, is that of the electrodeposited gold defining the CPW line and ground planes. Finally, the higher level, shown in red, is the elevated MEMS membrane. The central suspended plate with holes and the meandered suspensions are clearly visible. Any additional information about the technology is available in figure 4.2 and table 4.1.

Besides white light static 3D profiling, interferometric microscopy also enables measurements in dynamic conditions. In this case, the DUT is illuminated with stroboscopic (pulsed) light. This type of measurement can be effectively exploited just to observe periodic behavioural characteristics. The DUT is stimulated by a periodic waveform (e.g. sine, pulse, etc) at a certain frequency. On the other hand, the stroboscopic light is pulsed at the same frequency. The phase delay of the pulse feeding the light source with respect to the bias driving the DUT is increased, period after period. This way, at each iteration of the periodic behaviour, an image of the DUT in a different point of its characteristic is acquired. Finally, by putting together all the collected snapshots, the behaviour of the observed sample over an entire period is reconstructed [3–6].

The RF-MEMS sample in figure A.1 is biased with a zero mean value triangular waveform ranging between −20 V and +20 V. The absolute peak value is selected in order to be larger than the pull-in voltage. Moreover, its frequency is kept low (0.05 Hz) in order to perform the measurement in quasi-static conditions. The bias waveform versus time, with markers in correspondence of the pull-in and pull-out transitions, is reported in figure A.2.

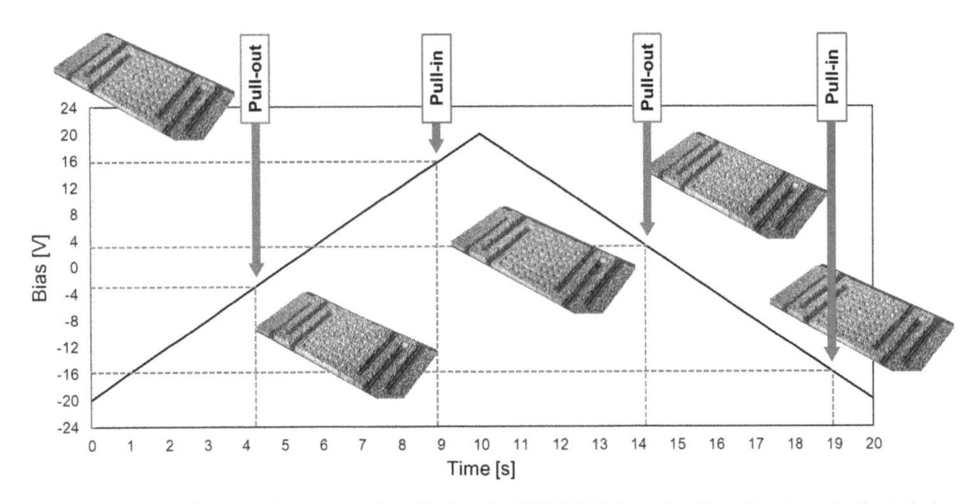

Figure A.2. Triangular waveform versus time biasing the DUT. Pull-in and pull-out levels are indicated along with measured images of the RF-MEMS when actuated (membrane colour closer to green) and not actuated (membrane colour closer to red).

Video A.1. Video animation of the RF-MEMS in figure A.1 when biased by the waveform in figure A.2, as reconstructed by 3D stroboscopic light optical dynamic interferometry.

Pull-in takes place at ±16 V, while pull-out happens at ±3 V. With the waveform centred around zero, two actuations and releases are observed over one period. The small measured snapshots of the RF-MEMS movable membrane help one to understand what it looks like (in terms of colours) when actuated and not actuated. Finally, the animation of the RF-MEMS, as reconstructed by the 3D stroboscopic light optical interferometric systems, is reported in Video A.1.

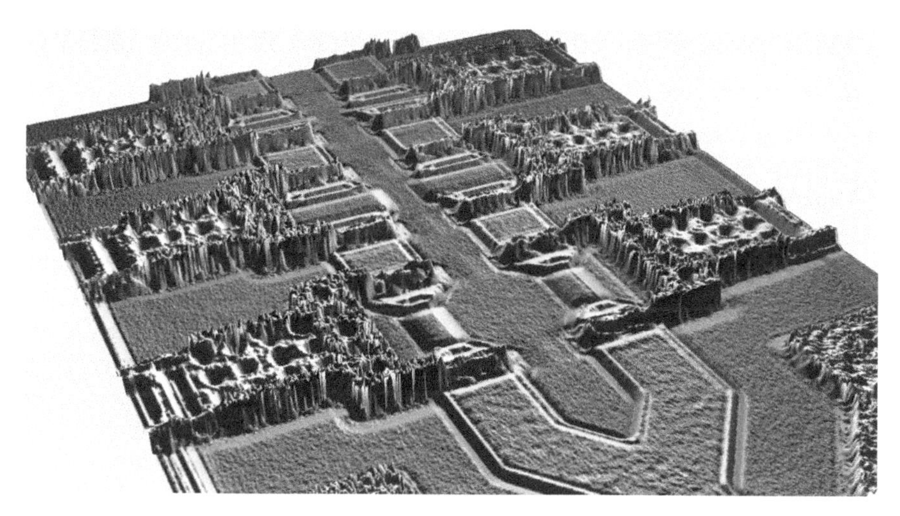

Figure A.3. 3D image of the RF-MEMS power attenuator discussed in [7–8], obtained by means of optical interferometry. The colour scale indicates the vertical quote of the DUT.

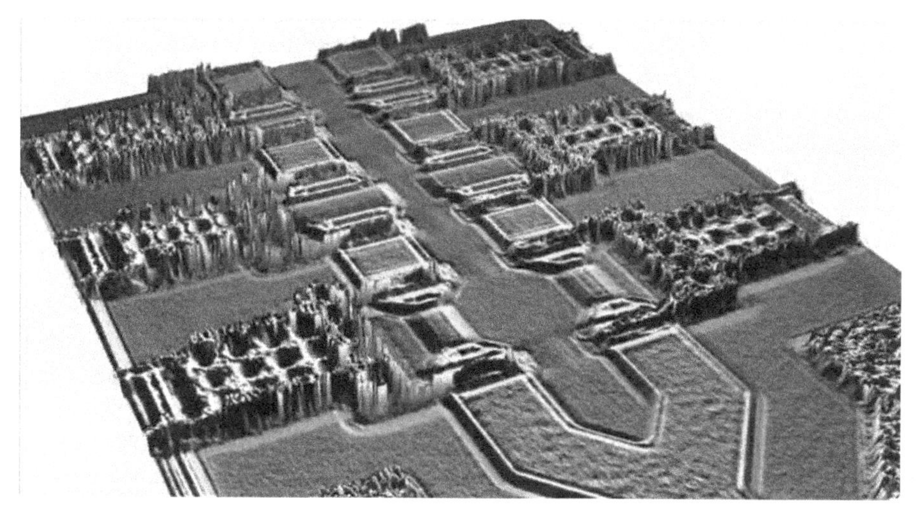

Video A.2. Video animation of the RF-MEMS attenuator in figure A.3 when biased by the waveform similar to that in figure A.2, as reconstructed by 3D stroboscopic light optical dynamic interferometry.

The dynamic characterization performed on a different DUT by means of the same measurement procedure discussed above is going to be reported. The observed RF-MEMS device is a multi-state power attenuator comprising six cantilevered ohmic switches, exploited to select or short resistive loads on a split RF line [7–8]. The 3D topology of the physical sample, obtained by means of white light interferometry, is shown in figure A.3.

The DUT is stimulated with a waveform similar to that previously discussed in figure A.2. Transitions of cantilevered switches, in correspondence to pull-in and pull-out, are visible in the animation, reconstructed by the 3D stroboscopic light optical interferometric systems, in Video A.2.

A.3 Measurements based on Laser Doppler Vibrometer (LDV) microscopy

In this section a different measurement technique, able to capture the dynamic characteristics of RF-MEMS, is exploited. It is called Laser Doppler Vibrometer (LDV) microscopy, and exploits the Doppler effect in order to reconstruct the movement of a certain spot pointed by a laser light source, i.e. if getting closer or farther with respect to the source [9–11]. Differing from dynamic interferometry, LDV microscopy enables one to capture the real-time evolution of the DUT. In other words, there is no need to cycle the sample many times to build its dynamics over one period. Nonetheless, the laser spot monitors just one point over the DUT surface. Therefore, if one wants to observe the dynamic characteristic of a certain sample portion, the biasing waveform has to be cycled several times, in order for the laser to acquire a different spot, period after period, and then reconstruct the evolution of a surface rather than of a single point.

The DUT example selected for LDV microscopy investigation is an RF-MEMS toggle-type switch. This particular geometry employs two suspended electrostatic actuators, anchored in their central point. Each actuator is provided with two distinct counter-electrodes for DC/AC biasing. Depending on whether the inner or outer electrodes are driven (pull-down and pull-up, respectively), the central plate can be displaced downwards or upwards, thus extending the vertical range it can span [12–14]. The layout of the RF-MEMS toggle-switch exploited for LDV microscopy dynamic measurement is shown in figure A.4.

In the experimental setup, the pull-down and pull-up electrodes are alternately biased, so that the actuators can tilt both clockwise and counter clockwise around the axis passing through flexible suspensions and anchoring points. The dynamic response captured by means of LDV microscopy is reported in Video A.3. In the animation, the surface measured with the laser source (i.e. the moving one) is reported in colours and the colour scale indicates the vertical displacement extent. Such a surface is superimposed to the grayscale measurement of the DUT. Since the entire RF-MEMS toggle-switch is larger than the maximum surface observable with the experimental setup, it was not possible to display the whole structure in the animation.

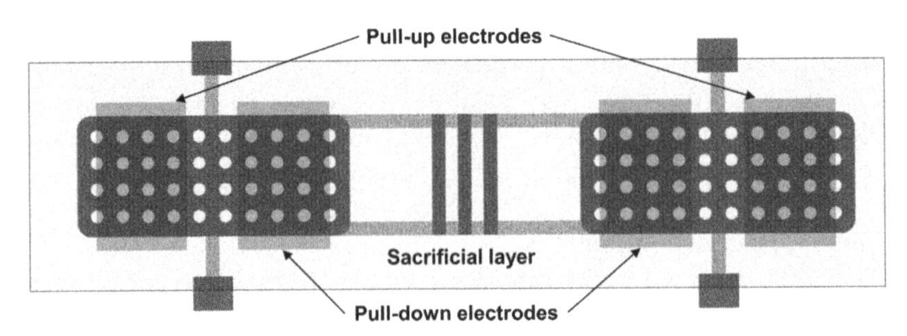

Figure A.4. Layout of the RF-MEMS toggle switch exploited for dynamic characterization based on LDV microscopy.

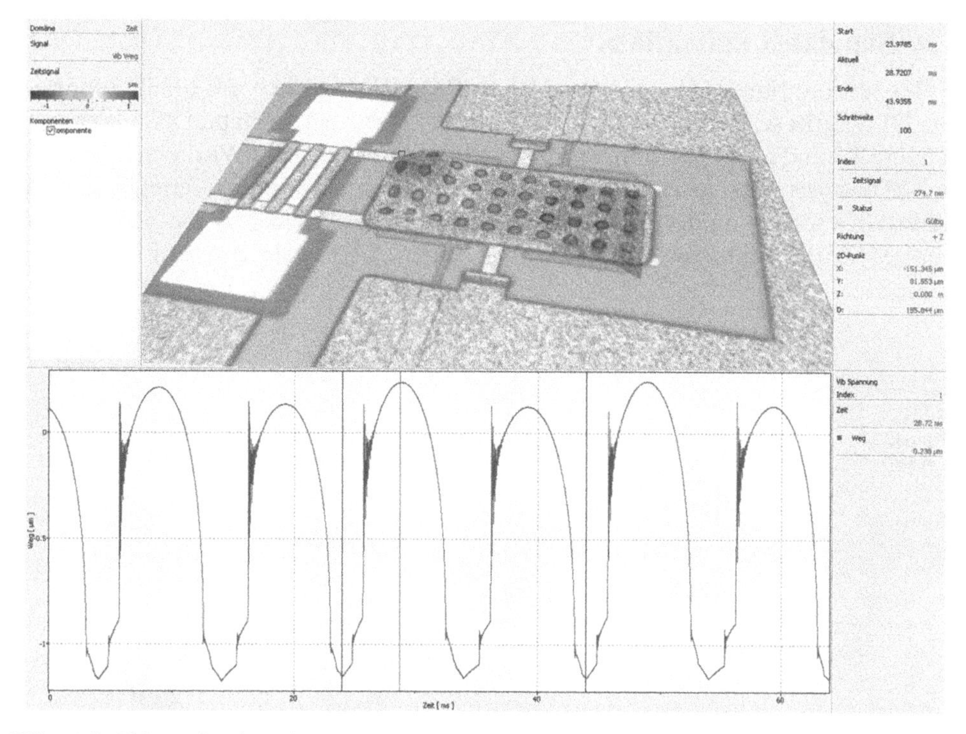

Video A.3. Video animation of the RF-MEMS toggle-switch in figure A.4 when pull-down and pull-up electrodes are alternately biased, as reconstructed by LDV microscopy dynamic measurement.

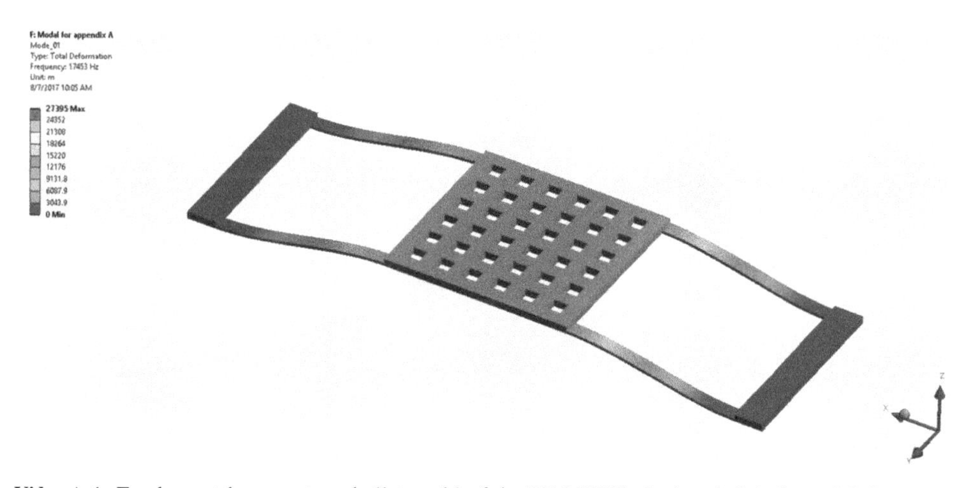

Video A.4. Fundamental resonant mode (1st mode) of the RF-MEMS ohmic switch in figure 4.5. Resonance is predicted by Workbench at 17.4 kHz.

A.4 Simulated animations

In this section, simulated animations of an RF-MEMS switch are briefly reported. The DUT is the series ohmic micro-relay already discussed in chapter 4 and reported in figure 4.3 and figure 4.5. Simulations are performed in Ansys Workbench, and the type of analysis is modal eigenfrequency, previously mentioned in section 4.3.2. The predicted resonant modes, from the fundamental (1st mode) to the 4th one, are reported in the animations in Video A.4, Video A.5, Video A.6 and Video A.7, respectively.

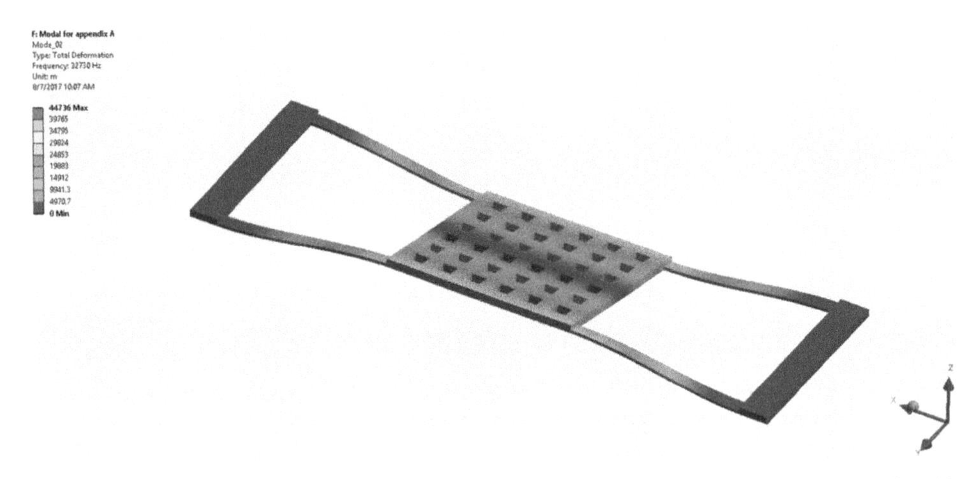

Video A.5. 2nd resonant mode of the RF-MEMS ohmic switch in figure 4.5. Resonance is predicted by Workbench at 32.7 kHz.

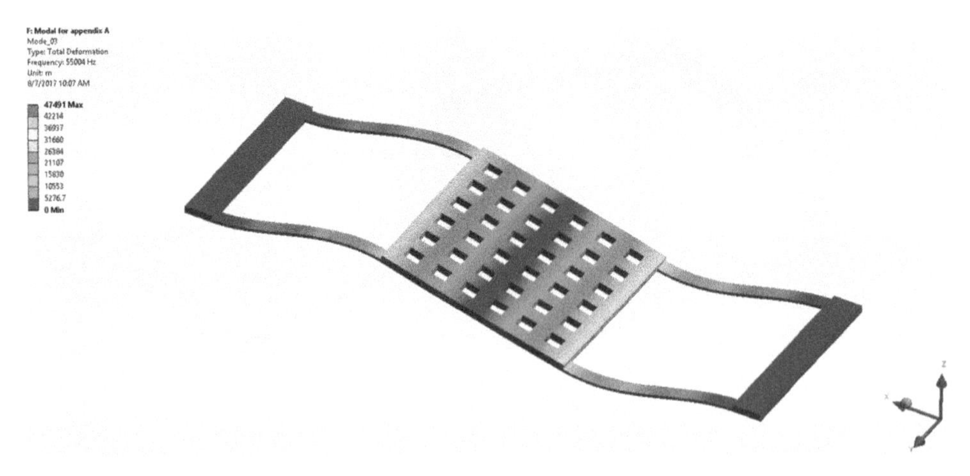

Video A.6. 3rd resonant mode of the RF-MEMS ohmic switch in figure 4.5. Resonance is predicted by Workbench at 55 kHz.

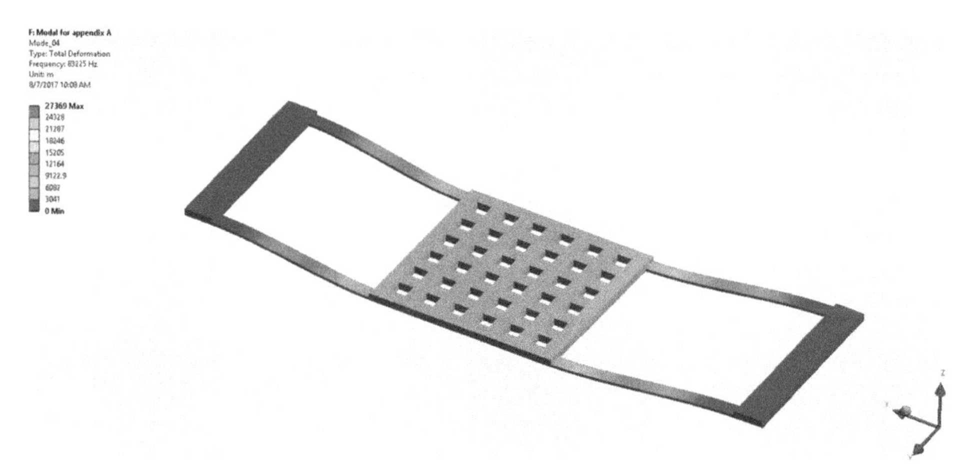

Video A.7. 4th resonant mode of the RF-MEMS ohmic switch in figure 4.5. Resonance is predicted by Workbench at 83.2 kHz.

Conclusion

In this appendix, a few significant examples of moving MEMS were reported. The reviewed data come both from experimental measurements of physical samples, as well as from Finite Element Method (FEM) simulations. Before showing any animation depicting Devices Under Test (DUTs) animated dynamics, the specific measurement/simulation procedure was described, in order to ease understanding of the presented material. The content of this appendix complements the overall discussion developed in the previous chapter of this work.

References

[1] Wang S H, Quan C, Tay C J, Reading I and Fang Z P 2003 Deformation measurement of MEMS components using optical interferometry *Meas. Sci. Technol.* **14** 909–15

[2] Blackshire J L and Sathish S 2002 Characterization of MEMS transducer performance using near-field scanning interferometry *IEEE Trans. Ultrason. Ferroelectr. Freq. Control* **49** 669–74

[3] Kim B, Razavi H A, Degertekin F L and Kurfess T R 2002 Micromachined interferometer for measuring dynamic response of microstructures *Proc. ASME Int. Mech. Eng. Congress and Exposition, MEMS Symp. (New Orleans, LA, Nov. 2002)*

[4] Graebner J E 2000 Optical scanning interferometer for dynamic imaging of high-frequency surface motion *Proc. IEEE Ultrasonics Symp. (San Juan, Oct. 2000)* pp 733–6

[5] Bosseboeuf A and Petitgrand S 2003 Characterization of the static and dynamic behaviour of M(O)EMS by optical techniques: Status and trends *J. Micromech. Microeng.* **13** 23–33

[6] Chen L-C, Huang Y-T and Chang P-B 2006 High-bandwidth dynamic full-field profilometry for nano-scale characterization of MEMS *J. Phys.: Conf. Ser.* **48** 1058–62

[7] Iannacci J, Faes A, Mastri F, Masotti D and Rizzoli V 2010 A MEMS-based wide-band multi-state power attenuator for radio frequency and microwave applications *Proc. of TechConnect World, NSTI-Nanotech (Anaheim, CA, June 2010)* pp 328–31

[8] Iannacci J, Faes A, Kuenzig T, Niessner M and Wachutka G 2011 Electromechanical and electromagnetic simulation of RF-MEMS complex networks based on compact modeling approach *Proc. of TechConnect World, NSTI Nanotech 2011 (Boston, MA, June 2011)* pp 591–4

[9] Lawrence E *Optical Measurement Techniques for Dynamic Characterization of MEMS Devices* http://www.polytec.com/fileadmin/user_uploads/Applications/Micro_Nano_Technology/Documents/OM_TP_MEMS_Whitepaper_2012_07_E.pdf [accessed 24 August 2017]

[10] Kim M G, Jo K, Kwon H S, Jang W, Park Y and Lee J H 2009 Fiber-optic laser Doppler vibrometer to dynamically measure MEMS actuator with in-plane motion *IEEE J. Microelectromech. Syst.* **18** 1365–70

[11] Liu J-W, Huang Q-A, Song J and Tang J-Y 2008 Detection of stiction of suspending structures in MEMS by a laser Doppler vibrometer systems *Proc. Int. Conf. on Solid-State and Integrated-Circuit Technology (Beijing, Oct. 2008)* pp 2472–5

[12] Schauwecker B, Strohm K A, Simon W, Mehner J and Luy J F 2002 Toggle-switch—a new type of RF MEMS switch for power applications *Proc. IEEE MTT-S Int. Microw. Symp. Digest (Seattle, WA, June 2002)* pp 219–22

[13] Solazzi F, Tazzoli A, Farinelli P, Faes A, Mulloni V, Meneghesso G and Margesin B 2010 Active recovering mechanism for high performance RF MEMS redundancy switches *Proc. Europ. Microw. Conf. (Paris, Sept. 2010)* pp 93–6

[14] Schauwecker B, Strohm K M, Mack T, Simon W and Luy J F 2003 Single-pole-double-throw switch based on toggle switch *IET Electron. Lett.* **39** 668–70

Lightning Source UK Ltd.
Milton Keynes UK
UKHW05n1639230518
323082UK00003B/92/P

9 780750 315432